SpringerBriefs in Biochemistry and Molecular Biology

More information about this series at http://www.springer.com/series/10196

Zhu Zeng · Xiaofeng Xu · Dan Chen

Dendritic Cells: Biophysics, Tumor Microenvironment and Chinese Traditional Medicine

 Springer

Zhu Zeng
School of Biology and Engineering
Guizhou Medical University
Guiyang, Guizhou
China

Dan Chen
Department of Pharmacology
School of Basic Medical Sciences
Tianjin Medical University
Tianjin
China

Xiaofeng Xu
Department of Laboratory Medicine
Beijing Tongren Hospital
Capital Medical University
Beijing
China

ISSN 2211-9353   ISSN 2211-9361   (electronic)
SpringerBriefs in Biochemistry and Molecular Biology
ISBN 978-94-017-7403-1   ISBN 978-94-017-7405-5   (eBook)
DOI 10.1007/978-94-017-7405-5

Library of Congress Control Number: 2015950874

Springer Dordrecht Heidelberg New York London

Springer Science+Business Media B.V. Dordrecht is part of Springer Science+Business Media (www.springer.com)

Contents

Abstract

Dendritic cells (DCs) are potent and specialized antigen-presenting cells, which play a crucial role in initiating and amplifying both the innate and adaptive immune responses. Functionally, DCs exist in two differentiation stages: immature DCs (imDCs) and mature DCs (mDCs). The imDCs are present in non-lymphoid tissues. Upon capturing antigens, they travel via the blood or lymph to secondary lymphoid organs, and gradually differentiate into mDCs which upregulate the expressions of peptide–MHC complexes and accessory molecules (CD11c, CD80, CD83, CD86, CCR7, etc.) on their surfaces that are necessary for naive T cell activation, leading to immune response or tolerance. Any interaction information between DCs and naive T cells could be translated into a new clinical therapy against diseases. Clinically, the DCs-based immunotherapy against cancer (cancer vaccination) has achieved some successes and is considered one of the most promising therapies to overcome cancers, but there are still many conundrums need to be solved. The motility of DCs is especially crucial for migration of imDCs into peripheral tissue and physical interaction between mDCs and naive T cells in the secondary lymph node. The biophysical characteristics of cells can reflect their relationship between structures and functions. Therefore, this book concentrates on the investigations of DCs at different differentiation stages and under various tumor microenvironments, as well as Chinese traditional herbs-conditioned media from the interdisciplinary viewpoints of biophysics, tumor immunology, and cell biology. The results showed that the DCs at different differentiation stages appear with various biophysical characteristics, including osmotic fragility, electrophoretic mobility, deformability, membrane fluidity, F-actin organization, and Fourier-transformed infrared spectra. The tumor microenvironment-derived factors (TMDF) impair the biophysical properties of DCs, one of the Chinese traditional herbs can partly recover the impairments in the biophysical characteristics of DCs, moreover, these changes are closely correlated with the expression levels of some cytoskeleton-binding proteins. It is significant for further understanding of the biological behaviors of DCs and the immune escape mechanism of cancer, as well as how to enhance the effectiveness of the DCs-based immunotherapy against cancers.

Chapter 1
Introduction

Abstract Since the discovery of DCs by Ralph M. Steinman, a lot of great progresses have been achieved leading to further understanding of immune response and their clinical application to therapies of immune-associated diseases. Here, some medical implications of DC's biology are reviewed in order to account for illness and provide opportunities for prevention and therapy.

Keywords DCs · Biology · Immune response · Diseases · Cancer

DCs are specialized and professional antigen-presenting cells, which have potent ability to capture and process antigens and present tumor-associated antigens to naive T cells, therefore, induce antigen-specific immune response [1, 2]. DCs originate from bone marrow, which are positioned in all tissues of body [1–3]. DCs are ready for sampling the milieu and to transmit the gathered information to adaptive immune system [1–3]. DCs initiate an immune response through the presentation of the captured antigen, which is in the form of peptide–major histocompatibility complex (MHC) molecule complexes, to naive T cells and B cells in lymph node [1–3]. Mice and humans have two major subsets of DCs: myeloid DCs (also known as conventional DCs or classical DCs) and plasmacytoid DCs (pDCs) [4]. Functionally, DCs exist in two differentiation stages: imDCs and mDCs. The imDCs are present in peripheral tissues, and upon capturing antigens they subsequently migrate via the blood or lymph vessels to secondary lymphoid organs; meanwhile, imDCs differentiate into mDCs which upregulate the expressions of MHC complexes and accessory molecules, including CD11c, CD80, CD83, CD86, and CCR7 on their surfaces that are necessary for the physical interaction between naive T cells and mDCs in lymph node, thus, initiating an immune response [2, 5]. The imDCs can present self-antigens to naive T cells, resulting in immune tolerance either via T cell deletion or via the differentiation of regulatory or suppressor T cells. On the other hand, the mDCs can initiate the differentiation of antigen-specific T cells into effector T cells with special functions and cytokine profiles. The lymph node-resident DCs that acquired antigen directly from the lymph node first present peptides to naive $CD4^+$ T cells, which leads to T cell activation and interleukin-2

© The Author(s) 2015
Z. Zeng et al., *Dendritic Cells: Biophysics, Tumor Microenvironment and Chinese Traditional Medicine*, SpringerBriefs in Biochemistry and Molecular Biology, DOI 10.1007/978-94-017-7405-5_1

1

(IL-2) production that in turn facilitates T cell proliferation and clonal expansion. Subsequently, tissue-resident DCs that captured antigen in peripheral tissues migrate into the lymph node and present peptides to the already activated CD4$^+$ T cells, which facilitates the generation of effector T cells. On interaction with DCs, naive CD4$^+$ T cells and CD8$^+$ T cells can differentiate into antigen-specific effector T cells with different functions. CD4$^+$ T cells can become T helper 1 (Th$_1$) cells, Th$_2$ cells, Th$_{17}$ cells, or T follicular helper (T$_{FH}$) cells that help B cells to differentiate into antibody-secreting cells, as well as regulatory T (T$_{Reg}$) cells that downregulate the functions of other lymphocytes. Naive CD8$^+$ T cells can give rise to effector cytotoxic T lymphocytes (CTLs). The types of T cell response, e.g., CD4$^+$ helper T cells or CD8$^+$ CTLs, are at least partly linked to the subset of DCs that presents the antigen. DCs can also interact with cells of the innate immune system, including natural killer (NK) cells, phagocytes, and mast cells [4, 6]. In addition, DCs also have an important role in controlling humoral immunity, which, respectively, directly interact with B cells and indirectly promote the expansion and differentiation of CD4$^+$ helper T cells [7, 8]. The above-mentioned important characteristics of DCs, which promote the activation of both arms of the adaptive immune system (cellular and humoral immune response) and which launch the immune response, render DCs the central candidates for antigen delivery and therapeutic vaccination against cancer [4]. Clinically, any interaction information between DCs and T cell in lymph node could develop a new therapy against diseases [1].

In 1973, Ralph M. Steinman first discovered a novel cell type from mouse spleen, which was named DCs for their distinct morphological features [9]. DCs were demonstrated as potent inducers of the "mixed leukocyte reaction" in mice until 1978 [10]. Then, the observation of high levels of MHC molecules on DCs was published by Steinman and his companies after one year [11]. The first human DCs were characterized in peripheral blood by Van Voorhis et al. [12], but their low numbers in blood circulation led to the characterization of DCs populations in humans challenging and the generation of DCs was so difficult in vitro that studies on DCs were at low ebb in the 1980s, following that, imDCs and mDCs were identified in late 1980s [13, 14]. It was not until 1992 that Steinman's group successfully generated large numbers of DCs from mouse, two years later; large numbers of DCs could be mobilized by specific cytokines from progenitors leading to a boom in DCs studies [14–16]. From then, a lot of DCs-based vaccination against cancer were developed and translated into clinical trials [1–4, 6, 17, 18]. The first DCs vaccine was approved by the U.S. Food and Drug Administration in 2010. DCs are really on the main stage of immunotherapy [19, 20]. Presently, the DCs-based cancer vaccination ex vivo has achieved some successes and is considered one of the most promising therapies to cure cancers, but there are still many challenges needed to be overcome, e.g., antigen selection, injection pathway and interval, immune escape mechanism of tumor, etc, especially, the overall number of ex vivo DCs reaching a lymph node is very small (probably less than 1 %) after their intracutaneous injection into a host loading tumor [5, 21–24]. Nowadays, a mass of strategies is focused on the further understanding of the biological behaviors of DCs and how to acquire the best clinical efficacy of DCs-based immunotherapy against tumor [1, 2, 6, 25]. Therefore, the elucidation of biological characteristics of DCs under pathological conditions (especially tumors) from biophysical viewpoint is an urgent research mission. In this book, we summarize the results about DC's biophysical studies in order to provide fresh ideas about DC's biology and DCs-based immune vaccination.

References

1. Steinman RM, Banchereau J. Taking dendritic cells into medicine. Nature. 2007;449 (7161):419–26.
2. Steinman RM. Decisions about dendritic cells: past, present, and future. Annu Rev Immunol. 2012;30:1–22.
3. Banchereau J, Steinman RM. Dendritic cells and the control of immunity. Nature. 1998;392 (6673):245–52.
4. Palucka K, Banchereau J. Cancer immunotherapy via dendritic cells. Nat Rev Cancer. 2012;12 (4):265–77.
5. Schreiber RD, Old LJ, Smyth MJ. Cancer immunoediting: integrating immunity's roles in cancer suppression and promotion. Science. 2011;331(6024):1565–70.
6. Sabado RL, Bhardwaj N. Dendritic cell immunotherapy. Ann NY Acad Sci. 2013;1284: 31–45.
7. Jego G, et al. Dendritic cells control B cell growth and differentiation. Curr Dir Autoimmun. 2005;8:124–39.
8. Qi H, et al. Extrafollicular activation of lymph node B cells by antigen-bearing dendritic cells. Science. 2006;312(5780):1672–6.
9. Steinman RM, Cohn ZA. Identification of a novel cell type in peripheral lymphoid organs of mice. I. Morphology, quantitation, tissue distribution. J Exp Med. 1973;137(5):1142–62.
10. Steinman RM, Witmer MD. Lymphoid dendritic cells are potent stimulators of the primary mixed leukocyte reaction in mice. Proc Natl Acad Sci USA. 1978;75(10):5132–6.
11. Steinman RM, et al. Identification of a novel cell type in peripheral lymphoid organs of mice. V. Purification of spleen dendritic cells, new surface markers, and maintenance in vitro. J Exp Med. 1979;149(1):1–16.
12. Van Voorhis WC, et al. Human dendritic cells. Enrichment and characterization from peripheral blood. J Exp Med. 1982;155(4):1172–87.
13. Witmer-Pack MD, et al. Granulocyte/macrophage colony-stimulating factor is essential for the viability and function of cultured murine epidermal Langerhans cells. J Exp Med. 1987;166 (5):1484–98.
14. Romani N, et al. Presentation of exogenous protein antigens by dendritic cells to T cell clones. Intact protein is presented best by immature, epidermal Langerhans cells. J Exp Med. 1989;169(3):1169–78.
15. Inaba K, et al. Generation of large numbers of dendritic cells from mouse bone marrow cultures supplemented with granulocyte/macrophage colony-stimulating factor. J Exp Med. 1992;176(6):1693–702.
16. Romani N, et al. Proliferating dendritic cell progenitors in human blood. J Exp Med. 1994;180 (1):83–93.
17. Nestle FO. Dendritic cell vaccination for cancer therapy. Oncogene. 2000;19(56):6673–9.
18. Cintolo JA, et al. Dendritic cell-based vaccines: barriers and opportunities. Future Oncol. 2012;8(10):1273–99.

19. Kantoff PW, et al. Sipuleucel-T immunotherapy for castration-resistant prostate cancer. N Engl J Med. 2010;363(5):411–22.
20. Kawalec P, et al. Sipuleucel-T immunotherapy for castration-resistant prostate cancer. A systematic review and meta-analysis. Arch Med Sci. 2012;8(5):767–75.
21. Randolph GJ, Angeli V, Swartz MA. Dendritic-cell trafficking to lymph nodes through lymphatic vessels. Nat Rev Immunol. 2005;5(8):617–28.
22. Dunn GP, et al. Cancer immunoediting: from immunosurveillance to tumor escape. Nat Immunol. 2002;3(11):991–8.
23. Zitvogel L, Tesniere A, Kroemer G. Cancer despite immunosurveillance: immunoselection and immunosubversion. Nat Rev Immunol. 2006;6(10):715–27.
24. Zou W. Immunosuppressive networks in the tumour environment and their therapeutic relevance. Nat Rev Cancer. 2005;5(4):263–74.
25. Benencia F, et al. Dendritic cells the tumor microenvironment and the challenges for an effective antitumor vaccination. J Biomed Biotechnol. 2012;2012:425476.

Chapter 2
The Measurement Protocols for Cell's Biophysical Characteristics

Abstract The biophysical characteristics of cells can reflect the relationship between structure and function. It is significant for understanding the physiological function of cells. Here, some measure protocols are described in detail.

Keywords Fluorescence polarization · Osmotic fragility · Deform capability · Adhesion · Electrophoretic mobility · FTIR

2.1 Isolation of Monocytes (MOs) and Generation of DCs

DCs were generated from fresh peripheral blood mononuclear cells of healthy human subjects, who had given the informed consent for the experimental study, which was approved by the Ethics Committee of Health Science Center of Peking University, as described by Steinman [1] with minor modification. In brief, $CD14^+$ MOs were obtained after plastic adherence and purified to about 97 % by depletion of non-$CD14^+$ cells by using cocktail immunomagnetic beads as described by the manufacturer (Miltenyi Biotec, Bergisch, Germany). The cells were then cultured in RPMI 1640 medium-10 % fetal bovine serum supplemented with 50 ng/mL recombinant human granulocyte-macrophage colony stimulating factor (rhGM-CSF, R&D Systems) and 1000 U/mL rhIL-4 (R&D Systems, Minneapolis, MN) for 7 d to develop into imDCs. Maturation was induced by addition of 100 U/mL rhTNF-α (R&D Systems) to imDCs for another 3 d of culture. Monocytes, imDCs, and mDCs were harvested for all measurements. The phenotypes of these three types of cells were analyzed by cell surface staining by using fluorescein isothiocyanate or phycoerythrin-conjugated mouse antibodies (Sigma, St. Louis, MO) to human CD1a, CD14, CD11c, CD80, CD86, CD40, CCR7, and HLA-DR with a flow cytometer (FACScan; BD Biosciences, San Jose, CA). To measure the viability of different cells, routine trypan blue staining was performed as routine protocol. Conventional scanning electron microscopy (JSM-5600LV: JOEL, Tokyo, Japan) was performed to obtain morphological images of the cells.

2.2 Fluorescence Polarization

Washed cells (2×10^6 cells/mL) in phosphate-buffered saline (PBS) were incubated at room temperature with an equal volume of 2×10^6 M 1, 6-diphenyl-1, 3, 5-hexatriene (DPH) in PBS for 30 min. Control cells were treated identically, but with the omission of the labeling reagent DPH. The final concentration of the solvent tetrahydrofuran in which DPH had been predissolved was 0.05 %. Excess reagent was removed by centrifugation at 600 g for 5 min, and the cells were resuspended in PBS. Steady-state fluorescence was measured with a fluorescence spectrometer (Hitachi, Tokyo, Japan). The fluorescence polarization parameter P was determined according to Azumi [2]. Wavelengths of 360 and 430 nm were used for excitation and emission, respectively. Measurements were completed within 2–3 h after harvesting. Each experiment was performed at least in triplicate.

2.3 Cell Osmotic Fragility

Solutions with 10 different osmotic concentrations (25–295 mOsm/kg) were prepared by mixing PBS with distilled water in various proportions. Ten aliquots of 1-mL cell suspensions (2×10^6 cells/mL) were placed in 10 centrifuge tubes. The supernatant was removed after centrifugation at 600 g for 5 min. Then, 500 μL of solution with different osmotic concentrations was added to each of these tubes and well mixed. Thirty minutes later, the numbers of nonhemolyzed cells in different tubes were counted with a blood cell counting chamber, and the number was divided by that in the control tube (295 mOsm/kg) to obtain the percentages of nonhemolyzed cells.

2.4 Cell Deformcapability

As shown in Fig. 2.1, the micropipette aspiration system was composed of an inverted microscope, a micromanipulator, a video recorder, a pressure control and recorder system, and a pipette with an internal diameter of 2.4–3.1 μm. Cell suspension (0.5 mL, 1×10^6 cells/mL) was infused into a chamber located on the specimen stage of the microscope. The pipette tip was positioned near the surface of a cell by using a micromanipulator. Negative pressure was applied to the micropipette to aspirate a small portion of the cell into micropipette. The time course of cell deformation was continuously recorded on the video recorder. Sequential photographs were taken from the recorder video image during single-frame replay

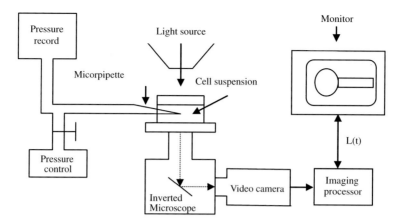

Fig. 2.1 Schematic diagram of the micropipette system

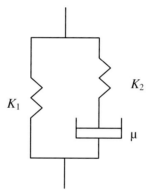

Fig. 2.2 Schematic drawing showing three viscoelastic elements used to model the rheological properties of monocytes, imDCs, and mDCs. K_1 and K_2 are elastic elements, and μ is a viscous element

on the video monitor every 120 ms (Fig. 2.2). The length of the cell tongue aspirated into the micropipette was determined as a function of time. The time history of deformation typically showed an initial rapid phase followed by a slow creep, similar to the behavior of peripheral blood leukocytes studied previously [3]. A standard solid viscoelastic model shown schematically in Fig. 2.3 was used to fit the experimental data. The equation of this model is as follows:

$$\sigma + (\mu/K_2)\partial\sigma/\partial t = K_1\varepsilon + \mu(K_1/K_2)\partial\varepsilon/\partial t$$

where $\partial\sigma/\partial t$ and $\partial\varepsilon/\partial t$ are stress and strain are partial derivatives of stress and strain as a function of time, respectively. K_1 and K_2 are elastic elements, and μ is a viscous element.

2.5 Flow Chamber Assay

A parallel plate flow chamber system [4] was used to assess the adhesion of MO and DC to endothelial monolayers. As shown in Fig. 2.3, the flow circuit was composed of a flow chamber, two reservoirs and a pump (Millipore Co., Bedford, USA). The two reservoirs were placed on the same altitude to remove the influence of static water pressure. The mean shear stress to which the cells were exposed was expressed as: $\tau = 6\eta Q/h^2$ w, where τ was the shear stress, η was the viscosity of physiological saline (0.707 cP), Q was the flow rate, h was the channel height (0.3 mm), and w was the channel width (1.0 cm). The desired shear stress τ was realized by adjusting the roller pump to attain different Q. The temperature was kept 37 °C all through the assay. A slide covered with a monolayer of HUVECs was placed in the flow chamber. 2×10^5 cells were injected into the chamber through a three-way tap. After 5 min adherence, physiological saline was pumped into the chamber at low shear stress (0.1 Pa) for 1 min to remove non-adherent cells. Thereafter, various shear stresses (0.5–6.5 Pa) were respectively imposed on the adherent cells for 2 min. The numbers of the adherent cells in ten random cell fields were counted, and the adhesion percentages were calculated and expressed as the percentage of the number of adherent cells after shearing in that before shearing (after pre-shearing at 0.1 Pa). Five repeated experiments were performed to acquire the mean value and standard deviation for each type of cells. For blocking assays,

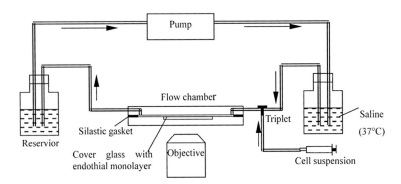

Fig. 2.3 Schematic of flow chamber system employed in the adhesion assay (cited from Clinical hemorheology and microcirculation, 2005, 32(4): 261–268)

as Jadhav et al. [5, 6] described, cells (10^7/mL) were incubated for 10 min at 37 °C with anti-CD11a MAb (20 µg/mL) before addition to the flow chamber.

2.6   Cell Electrophoretic Mobility

The cells were adjusted to 2×10^6/mL with 9 % saccharum solution. Their electrophoretic mobility (EPM, voltage 40 V, 30 °C) was examined by a cell electrophoresis meter (LIANG-100, Shanghai Medical University, China). Ten cells were randomly selected for a sample. Five repeated experiments were performed to obtain the means.

2.7   Fourier-Transformed Infrared Spectra

Cells (5×10^6) were washed twice in heavy water containing 0.9 % NaCl by centrifugation at 600 g for 5 min. The pellet was deposited on the CaF_2 window and evaporated for 10 min at 37 °C. A thin circular moist film with a diameter of 2–3 mm was formed. The window with the film was positioned in the NEXUS-470 Fourier-transform infrared (FTIR) spectrometer (Thermo Electron, Waltham, MA) range from 900 to 4000 cm^{-1}. The residual heavy water was interactively subtracted. All of the spectra were baseline-corrected and subjected to Fourier self-deconvolution. According to the results of the Ramesh group [7, 8], the absorption intensity ratios of $A_{1020/A1545}$, A_{1121}/A_{1545}, A_{1030}/A_{1080}, and A_{1030}/A_{2924}, respectively, correspond to DNA/amide II (relative content of DNA and proteins), RNA/amide II (transcriptional states), glucose/phosphate (metabolic turnover), and glucose/phospholipid (*de onvo* synthesis of phospholipids at the expense of free glucose). All spectra were subjected to Fourier self-deconvolution, and there was no wave number shift in the spectra (Fig. 2.4).

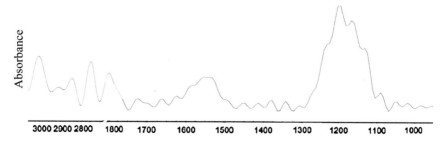

Fig. 2.4 The schematic diagram of infrared spectra of the contents of lipids and proteins in cells (cited from Cell Biochem Biophys. 2006, 45(1):19–30)

2.8   Confocal Laser Scanning Microscopy Analysis

Cells (2×10^6) were washed twice with PBS, pH 7.4. The sample was fixed in 3.7 % formaldehyde solution in PBS for 10 min at room temperature. After washing twice in PBS, the cells were resuspended in 0.1 % Triton X-100/PBS for 5 min at room temperature and then incubated in 1 % bovine serum albumin/PBS for 30 min to reduce the nonspecific background staining. After washing in PBS, the cells were labeled with 2 U of a fluorescent derivative of phalloidin (Rhodamine phalloidin; Molecular Probes, Eugene, OR) for 20 min in the dark. The cells were washed with PBS once and resuspended in 50 µL of PBS. A small chamber was made on the object glass with the use of double-faced adhesive tape. The suspension was dropped into the chamber, which was covered with a cover glass and mounted on the stage of the confocal laser Scanning microscopy (CLSM; Leica Lasertechnik, Heidelberg, Germany). A 568 nm laser was used for the excitation of rhodamine phalloidin. A 580-nm long-pass filter was placed in the fluorescence detection path. The fluorescence was collected using a ×100 oil immersion objective. A series of ×2 optically zoomed confocal sections, 1–2 µm apart, were scanned, with each image averaged by 16 line scans. The images obtained were three-dimensionally reconstructed with the software function of Leica. Then, the mean relative fluorescence intensities of nine cells at each stage were determined by using the other function of same software.

References

1. Steinman RM. The dendritic cell system and its role in immunogenicity. Annu Rev Immunol. 1991;9:271–96.
2. Azumi M. Fluorescence assay in biology and medicine. J Chem Phys. 1962;37:2413–20.
3. Schmid-Schonbein GW, et al. Passive mechanical properties of human leukocytes. Biophys J. 1981;36(1):243–56.
4. Yao W, et al. Low viscosity Ektacytometry and its validation tested by flow chamber. J Biomech. 2001;34(11):1501–9.
5. Jadhav S, Konstantopoulos K. Fluid shear- and time-dependent modulation of molecular interactions between PMNs and colon carcinomas. Am J Physiol Cell Physiol. 2002;283(4): C1133–43.
6. Jadhav S, Bochner BS, Konstantopoulos K. Hydrodynamic shear regulates the kinetics and receptor specificity of polymorphonuclear leukocyte-colon carcinoma cell adhesive interactions. J Immunol. 2001;167(10):5986–93.
7. Ramesh J, et al. Application of FTIR microscopy for the characterization of malignancy: H-ras transfected murine fibroblasts as an example. J Biochem Biophys Methods. 2001;50(1):33–42.
8. Salman A, et al. FTIR microspectroscopy of malignant fibroblasts transformed by mouse sarcoma virus. J Biochem Biophys Methods. 2003;55(2):141–53.

Chapter 3
Biophysical Characteristics of DCs at Different Differentiation Stages

Abstract DCs are the most powerful antigen-presenting cells, which play key roles in initiation and amplification of innate and adaptive immune response. They and their precursors undergo complex migration to perform their functions in vivo. To explore their immune functions from biophysical viewpoint, the biophysical characteristics of DCs at different differentiation stages were investigated in vitro, including osmotic fragility, electrophoretic mobility, deformability, membrane fluidity, adhesion capability, F-actin cytoskeleton organization, and FTIR spectra. The results showed that the DCs at various differentiation stages displayed distinct biophysical characteristics. It is significant for further understanding the mechanisms of the activation of immunological responses and the migration from peripheral tissue to secondary lymphoid organs.

Keywords DCs · Biophysical characteristics · Immune function · Immune response

DCs are highly motile cells whose whole differentiation process displays dynamic migration property [1]. The biophysical characteristics are closely associated with their capabilities and corresponding functions. Herein we have discussed the biophysical behaviors of the DCs at different developmental stages, included membrane fluidities, membrane viscoelastic properties, osmotic fragilities, adhension characteristics, important cell components, cytoskeleton (F-actin) organization, and FTIR spectra, with the goal of gaining a new insight into the basic biology and clinical applications of DCs.

Membrane fluidity is a biophysical parameter, which reflects molecular motion and arrangement of the hydrocarbon chain region in lipid bilayer and in turn membrane dynamics and cell deformability [2]. The fluid state of membrane is thought to play an important role in cellular functions and processes such as growth and differentiation [3]. DC's membrane fluidity was studied by determining the fluorescence parameter p which is negatively correlated with membrane fluidity. When monocytes (MOs) differentiated into mDCs, the fluorescence polarization parameter p value significantly decreased (Table 3.1; Fig. 3.1), suggesting that the membrane fluidity was significantly increased at the process. DCs could execute

© The Author(s) 2015
Z. Zeng et al., *Dendritic Cells: Biophysics, Tumor Microenvironment and Chinese Traditional Medicine*, SpringerBriefs in Biochemistry and Molecular Biology, DOI 10.1007/978-94-017-7405-5_3

Table 3.1 Fluorescence polarization of monocytes, imDCs, and mDCs (mean ± SD)

	Mo	imDCs	mDCs
P	0.836 ± 0.016	0.125 ± 0.022	0.062 ± 0.002

Comparison of imDCs with Mo: $^*p < 0.01$
Comparison of mDCs with imDCs and monocytes $^{**}p < 0.001$ (cited from Cell Biochem Biophys. 2006, 45(1):19–30)

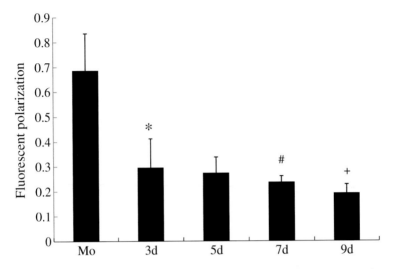

Fig. 3.1 Membrane fluidity of DCs at different stages. $^*p < 0.05$ compared with MOs; $^#p < 0.05$ compared with DCs on the fifth day (5d); $^+p < 0.05$ compared with DCs on the seventh day (7d) (cited from Clin Hemorheol Microcirc. 2010, 46(4):265–273)

their normal function only when the membrane fluidity improves with their differentiation from MOs. At the immature stage, DCs are located in nonlymphoid tissues and their major function, like "sentinel" is to take up and process the antigen. The increased mobility would be helpful for imDCs to patrol the body (including the peripheral capillaries, tissues, and organs) with the involvement of transvascular deformation. The increased fluidity in imDCs may promote their deformability and thus help to perform their sentinel function. Compared with MOs and imDCs, mDCs had the highest membrane fluidity, indicating that mDCs had the best motility. The main function of mDCs is to initiate antigen-specific immune responses. While naive T cells physically interact with DCs, the plasma membranes of T cells and DCs become positioned next to each other, allowing the reciprocal engagement of cell surface receptors, including the T-cell receptor-MHC-peptide complex and costimulatory molecules [4]. Changes in membrane fluidity may be associated with the mechanism of the variable exposure of surface receptors or ligands and antigen-peptide. The highest membrane fluidity of mDCs may result in the exposure of costimulated molecules (e.g., CD86, CD80) and signaling molecules (e.g., CD40), which could be recognized by T cells, thus mediating the activation of T cells.

The viscoelastic property of cells was identified by the micropipette experiment which showed that the elastic coefficients K_1 and K_2 significantly decreased with DC's differentiation (Table 3.2; Fig. 3.2). K_1 and K_2 determine the degrees of the initial rapid deformation and the maximum deformation, respectively [5–7]. Higher K_1 and K_2 values indicate lower deformability. That is, mDCs have the best deformabilities among the three types of cells. This deformability can be confirmed by the association with the expression level or the conformation changes of motor proteins such as myosins [8]. Also, the deformabilities may be related to the

Table 3.2 Viscoelastic coefficients of MOs, imDCs, and mDCs (mean ± SD)

Coefficient	MOs (n = 20)	imDCs (n = 20)	mDCs (n = 20)
K_1 (N/m^2)	30.5 ± 7.8	21.6 ± 3.6[*]	15.7 ± 2.9[**]
K_2 (N/m^2)	16.9 ± 4.2	11.3 ± 4.7[*]	8.3 ± 1.5[**]
μ (N m^2)	3.8 ± 2.7	4.5 ± 1.5	4.1 ± 1.5

Comparison of imDCs with MOs: [*]$p < 0.01$
Comparison of mDCs with mDCs: [**]$p < 0.01$ (cited from Cell Biochem Biophys. 2006, 45(1):19–30)

Fig. 3.2 Sequence of photographs that had been processed by Photoshop software for publication, ×400, showing the progressive deformation of MOs, imDCs, and mDCs in a micropipette in response to an aspiration pressure of 196 N/m^2. The deformability of mDCs is better than that of monocytes and imDCs ($p < 0.01$) (cited from Cell Biochem Biophys. 2006, 45 (1):19–30)

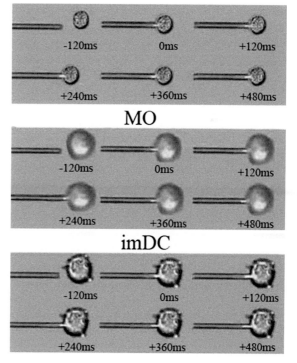

(a)   **(b)**

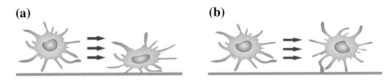

Fig. 3.3 Schematic diagram of the deformation of cells adherent to the vessel endothelial cells under the actions of blood flow shear force. The flow direction was from left to right. The cell with fine deformability flattened with significant adhesion to extracellular matrix or vascular endothelium, which could favor the transmigration (**a**), whereas the cell with poor deformability remained spherical and moved along the shear flow direction (**b**) (cited from Cell Biochem Biophys. 2006, 45(1):19–30 and revised for this publication)

changes in the expression level of cytoskeleton proteins [9]. Finer cell deformabilities contributed to a stronger adhesion by providing a larger contact area so that more adhesion or signaling receptor–ligand bonds could be formed. Moreover, if the cells were deformed by blood flow shear force, their decreased height would result in less vessel lumen obstruction and less force on cells (Fig. 3.3a). The better deformability also made cells easy to attach to the vascular endothelium or extracellular matrix and less susceptible to impairment by the flow shear force in vessels (Fig. 3.3b). Therefore, better deformability may function in concert with the increased adhesion to vascular endothelium or matrix and enhance the ability of imDCs and mDCs to emigrate from the sites of antigen uptake or circulation and then enter into second lymphoid tissues. Furthermore, the studies found that the greater membrane fluidity and deformability for mDCs than those of MOs and imDCs indicated that the mDCs are softer than MOs and imDCs [9]. In addition, in the process of conjugate formation of T cells with mDCs, the rigidity of T cells markedly decreased [10]. Therefore, it is possible that the T cells and mDCs may simultaneously become less rigid at the interaction phase, and this may be beneficial for the interaction of mDCs and naive T cells and the binding of super-molecular activation cluster.

The osmotic fragility of the cell reflects its ability to resist the stress imposed by different osmotic concentrations. Compared with imDCs, the osmotic fragility of mDCs was significantly lower (Fig. 3.4), suggesting the reduced ability of mDCs to withstand the decreased osmotic concentrations. In contrast, imDCs are able to adapt to the different microenvironments at various osmotic concentrations in vivo, consistent with their antigen uptake function.

The adhesion characteristics of cells are significantly associated with their migration in vivo. It was found that imDCs, mDCs, and MOs had different binding abilities to human umbilical vein endothelial cells (HUVECs) (Fig. 3.5), adhesion molecules (AM) expression (Fig. 3.6) and EPM (Figs. 3.7, and 3.8). Both MOs and DCs are blood cells, which are inevitably affected by the fluid shear stress of the flowing blood when they migrate into vessels. Physiologic shear stress varies depending on vessel diameters, for example, 0.43 Pa in aorta, 3.0 Pa in venules, and 3.7 Pa in capillaries [11]. The study shows the adherent behavior of DCs generated

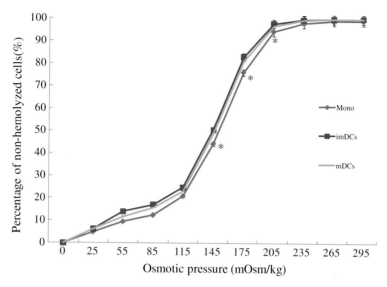

Fig. 3.4 Osmotic fragility curves for monocytes, imDCs, and mDCs. The percentage of nonhemolysed imDCs was significantly greater than that of the other two cell groups ($p < 0.05$) (cited from Cell Biochem Biophys. 2006, 45(1):19–30)

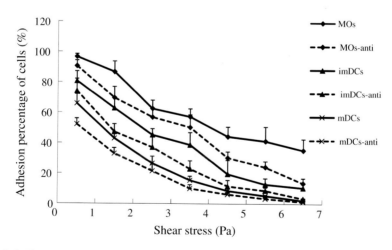

Fig. 3.5 The adhesion percentages of cells to HUVECs under various shear stresses in the flow chamber system. Under the same stress, the adhesion percentages decreased with the developmental stages of DCs ($p < 0.05$). After pre-incubation with anti-CD11a MAb, the percentages of adherent cells were all reduced ($p < 0.05$) (cited from Clinical hemorheology and microcirculation, 2005, 32(4): 261–268)

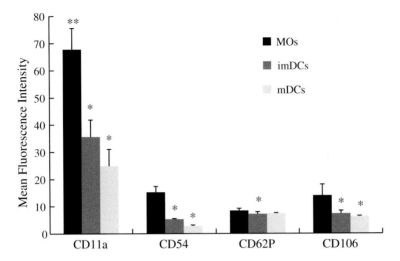

Fig. 3.6 Adhesion molecule expressions in MOs, imDCs, and mDCs examined by flowcytometric analysis. CD11a, CD54, CD62P, and CD106 expression were down-regulated gradually with the developmental stages of DCs ($^*p < 0.05$). The CD11a expression in the three types of cells was greatly different ($^{**}p < 0.01$). But CD54, CD62P, and CD106 expression in imDCs and mDCs were very low, and there was no significant difference between the two types of cells (cited from Clinical hemorheology and microcirculation, 2005, 32(4): 261–268)

from peripheral blood MOs to HUVECs under various shear stresses in a flow field, and mDCs had the weaker binding ability to HUVECs compared with MOs and imDCs (Fig. 3.5). The low binding ability of mDCs to vascular EC favored their migration with blood flow and interaction with T cells in lymph nodes. MOs had the highest binding ability to HUVECs in the three kinds of cells. MOs circulate in the blood and eventually traverse vascular endothelial lining to enter tissues, where they differentiate into macrophages or DCs [12]. Therefore, the firm adhesion of MOs to the endothelium would contribute to their emigration through the vessel wall [13]. AM include selectins, integrins, cadherins, immunoglobulin superfamily (IgSF), etc., and they play central roles in cell adhesion. CD11a is a subunit of leukocyte function-associated antigen (LFA-1, CD11a/CD18, and $\alpha_L\beta_2$), a member of integrin family that is predominantly involved in leukocyte trafficking and extravasation. LFA-1 is exclusively expressed in leukocytes and interacts with its ligands intercellular adhesion molecule (ICAM)-1, -2, and -3 to promote a variety of homotypic and heterotypic cell adhesion events required for normal and pathological functions of the immune systems [14]. Both CD54 (ICAM-1) and CD106 (vascular cell adhesion molecule-1, VCAM-1) are the structurally related members of the IgSF and are involved in cell adhesion, cell differentiation, lymphocyte migration and activation, as well as inflammation [15, 16]. CD62P (P-selectin) is mainly found in blood platelet and vascular EC and upregulates tissue factor in MOs and leads to leukocyte accumulation in areas of vascular injury associated with thrombosis and inflammation [17]. In vivo transmigration of MOs

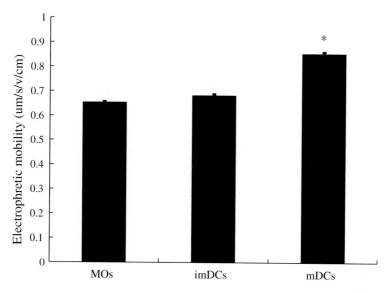

Fig. 3.7 The EPM of DCs at different differentiation stages. MOs and imDCs shared similar EPMs. But the EPM of mDCs increased significantly ($^*p < 0.01$) (cited from Clinical hemorheology and microcirculation, 2005, 32(4): 261–268)

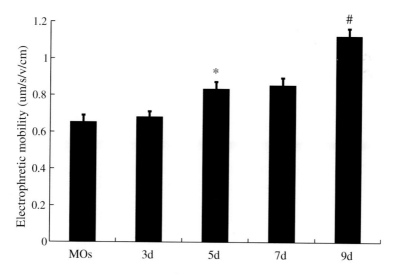

Fig. 3.8 EPM of DCs at different stages. $^*p < 0.05$ compared with DCs on the third day (3d); $^\#p < 0.05$ compared with DCs on the seventh day (7d) (cited from Clin Hemorheol Microcirc. 2010, 46(4):265–273)

is a multistep process that follows the well-known stepwise model of rolling, mediated by cell adhesion, involving integrins, and migration [18]. It could be speculated that migration of DCs from peripheral tissues to lymph nodes might follow the similar multistep process. During the maturation of DCs, the down-regulated CD11a, CD54, and CD106, especially CD11a, as detected from flow cytometric analysis (Fig. 3.6), might be related to their decreased adhesion to HUVECs. In order to test whether adhesion to HUVECs were mediated by CD11a, MOs and DCs were pre-incubated with anti-CD11a blocking antibody. The blockade resulted in a significant reduction in the adhesion percentages of cells, which supports the role of CD11a in mediating the adhesion of DCs and MOs to vascular EC. In addition, it had been reported that CD11b and CD36 were necessary for the binding of DCs to HUVECs, and both were drastically down-regulated during DCs maturation [19], which would also explain the reduced adhesion of mDCs. Cell adhesion is related not only to the AM, but also to the assembly of cytoskeleton. Siobhan Burns et al. had demonstrated that as maturation progressed, DCs became rounded and devoid of actin-rich structure known as podosomes. This change in morphology was closely associated with a quantitatively reduced ability to adhere to fibronectin or ICAM-1-coated surfaces [20]. Therefore, we speculated that the change in morphology also contributed to the decreased adhesion of mDCs to HUVECs. EPM is an important biophysical parameter, which reflects the amounts of cell membrane negative charges. Adhesion of cells to neighboring cells or to solid organic or inorganic surface has been suggested, for a long time, to be related to EPM [21]. It has been established that all cells of multicellular organisms have negatively charged surfaces as long as they live within their natural environments [22]. Under the same electric field, fast EPM of mDCs (Figs. 3.7 and 3.8) meant surplus negative charges on their surfaces compared with imDCs and MOs. Therefore, mDCs might suffer from powerful repulsive forces from negatively charged EC resulting in their binding being decreased. The study has found that the decrease of rabbit erythrocyte surface charges leads to the decrease in cell deformation, and impaired deformation of murine erythroleukemia cells makes their adhesion to HUVECs decrease [21, 23]. So the relevance to human DCs surface charges, cell deformation, and adhesion deserves further investigation. Besides the quantity of surplus negative charges, the cell's glycocalyces compactness is also an essential factor which had to be considered in interpreting the EPM data [24]. Compared with imDCs and MOs, mDCs might have less compact glycocalyces to be mechanically and hydrodynamically softer, so might exert less friction when migrating through an electrophoresis medium. Therefore, the adhesion ability of DCs in the flow field was dependent on their developmental stages, and CD11a and EMPs were involved in the adhesion.

The cytoskeleton (microfilaments) mainly control membrane plasticity (including cytoskeleton-propelled deformation and protrusion) and cell motility [25], regulate cell morphology through contraction and relaxation, and generate the mechanical force required for cell movement by association with motor proteins such as myosins [26]. The microfilament is the structural basis of DCs–T cell interactions. Compared with imDCs and mDCs, the microfilaments (composed of

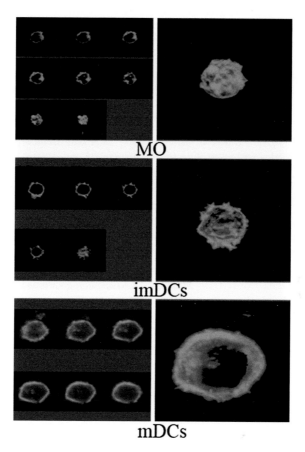

Fig. 3.9 Series of CLSM images (left) and three-dimensional images (*right*) of rhodamine phalloidin-labeled F-actin in monocytes, imDCs, and mDCs. After the three-dimensional reconstruction (*right*), it can be seen that F-actin was organized loosely and disorderly in imDCs and mDCs. The monocyte, in contrast, had a compact, conglobate, and symmetrical F-actin network. The F-actin cytoskeleton of imDCs and mDCs was asymmetric. The mean relative fluorescence intensity parameter I of F-actin of cell significantly decreased in imDCs and mDCs ($^*p < 0.01$ and $^{**}p < 0.001$, respectively) (cited from Cell Biochem Biophys. 2006, 45(1):19–30)

F-actin) in MOs were shorter and more compactly organized (Fig. 3.9). Furthermore, the F-actin expression levels of imDCs and mDCs were significantly lower than those of MOs (Table 3.3). The compact F-actin network might be involved in the decreased deformability and the increased osmotic fragility of MOs. With DC's differentiation, F-actin underwent the rearrangement and redistribution (Fig. 3.9). The microfilament reorganization in imDCs and mDCs may be related to the changes in its protein secondary structures. When MOs differentiate into DCs, the cells significantly remodel their shape and immunological dendritic pseudopods occur as a result of the cytoskeleton reorganization. The cytoskeletons of imDCs and mDCs were asymmetrical (Fig. 3.9). The characteristics of actin asymmetry

Table 3.3 The mean relative fluorescence intensity (I) of MOs, imDCs, and mDCs ($X \pm$ SD)

$N = 9$	MOs	imDCs	mDCs
I (mean relatively fluorescent intensity)	95.53 ± 1.75	79.72 ± 2.67[*]	37.09 ± 1.76[**]

Comparison of imDCs with monocytes: [*]$p < 0.001$
Comparison of mDCs with imDCs and MOs: [**]$p < 0.01$ (cited from Cell Biochem Biophys. 2006, 45(1):19–30)

and greater membrane fluidity of differentiating DCs may induce polarizations in membrane molecules and contribute to the changes of cell deformability.

Various groups have demonstrated the utility of FTIR microspectroscopy for elucidating the biochemical differences that distinguish the spectra of murine and human embryonic stem cells from their lineage-committed progeny. Heraud et al. showed that the differences between spectra of hESC lines and their derived cell types could be detected only after four days of differentiation [27]. Naumann et al. demonstrated that the FTIR spectrum of bacteria provides a unique fingerprint that allows the identification of bacteria species [28]. It was found that DCs at different differentiation stages (from CD14$^+$ MOs to mDCs) displayed various FTIR spectra [9].

As shown in Table 3.4, at the smaller wave numbers, such as 1020, 1030, 1080, and 1121 cm^{-1}, MOs have higher absorptions than imDCs and mDCs, indicating changes of DNA structures in imDCs and mDCs in comparison with MOs. At 1444 and 1465 cm^{-1}, the absorptions in imDCs are much higher than those in MOs and mDCs, suggesting that imDCs may have some changes in their membrane lipid composition, structure, or both. The components located at 1654, 1690, and 1545 cm^{-1} can be assigned to some given types of secondary protein structures (namely, random coil, turns, bends, and α-helix [29, 30]. Significant decreases in

Table 3.4 Mean absorptions at different wave numbers of FTIR spectra for MOs, imDCs, and mDCs (mean ± SD)

Wave number (cm^{-1})	MOs	imDCs	mDCs
1020	5.78 ± 1.33[*]	4.65 ± 0.96	0.51 ± 0.24
1030	5.63 ± 1.02[*]	4.33 ± 1.17	1.02 ± 0.19
1080	7.26 ± 0.88[*]	5.75 ± 1.28	1.07 ± 0.34
1121	5.70 ± 1.39[*]	5.37 ± 1.77	1.17 ± 0.56
1444	3.85 ± 0.71	9.78 ± 1.95[**]	2.08 ± 0.57
1465	5.52 ± 1.47	11.50 ± 2.68[**]	2.54 ± 0.93
1532	1.98 ± 0.32	1.11 ± 0.26	0.86 ± 0.12
1545	1.93 ± 0.29	1.22 ± 0.18	0.96 ± 0.14[***]
1645	6.28 ± 1.59	5.12 ± 1.46	1.88 ± 0.46[***]
1690	3.50 ± 0.93	1.57 ± 0.35	1.32 ± 0.24[***]
2924	4.42 ± 1.05	1.27 ± 0.13[*]	6.04 ± 1.66

Comparison of MOs with imDCs and mDCs [*]$p < 0.01$
Comparison of imDCs with MOs and mDCs [**]$p < 0.05$
Comparison of mDCs with MOs and imDCs [***]$p < 0.05$ (cited from Cell Biochem Biophys. 2006, 45(1):19–30)

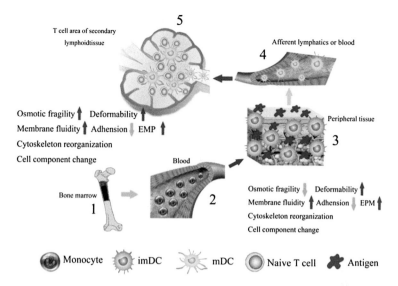

Fig. 3.10 Schematic drawing representing the migration course as DCs perform their physiological functions in vivo. From the first step to the fourth step, the cells undergo changes from MOs to immature and mature phases, which are correlated with the biophysical events involved in the transformations

the absorptions of imDCs and mDCs indicate that the contents of these protein secondary structures are reduced, which could lead to the reorganization of cytoskeleton. The imDCs have lower absorption at 2924 cm^{-1} than mDCs and MOs, suggesting that the membrane lipids of imDCs may have been altered through the differentiation process and this may be associated with the expression differences of lipid kinases that are related to the antigen uptake and processing. Above-mentioned changes could be the reasons that DCs at different differentiation stages possess various morphological characteristics and functions.

The DCs have been introduced as a therapy for tumors and autoimmune diseases before many years. However, the optimal immunization strategy is still not clear [31, 32]. Both the mature stage of DCs and the route of administration seem to be crucial. The mDCs are preferable to imDCs in immunization strategies for the biophysical property [32, 33]. The biophysical properties of DCs may be relevant to the execution of the functions of antigen uptake and processing, and T cell activation (summarized in Fig. 3.10). mDCs had maximal membrane fluidities, deformabilities, EPMs, less expression levels of some adhesion molecules, weaker adhesion to HUVECs, as well as a looser, disordered, and asymmetric cytoskeleton, various FTIR spectra, which will be helpful for clinicians to gain further insights into the scientific basis of DCs therapy against cancers and immune-related diseases.

References

1. Steinman RM. Dendritic cells: understanding immunogenicity. Eur J Immunol. 2007;37(Suppl 1):S53–60.
2. Nicolson GL. The fluid-mosaic model of membrane structure: still relevant to understanding the structure, function and dynamics of biological membranes after more than 40 years. Biochim Biophys Acta. 2014;1838(6):1451–66.
3. Nathan I, et al. Alterations in membrane lipid dynamics of leukemic cells undergoing growth arrest and differentiation: dependency on the inducing agent. Exp Cell Res. 1998;239(2):442–6.
4. Dustin ML. Making a little affinity go a long way: a topological view of LFA-1 regulation. Cell Adhes Commun. 1998;6(2–3):255–62.
5. Dong C, et al. Passive deformation analysis of human leukocytes. J Biomech Eng. 1988;110 (1):27–36.
6. Wu ZZ, et al. Comparison of the viscoelastic properties of normal hepatocytes and hepatocellular carcinoma cells under cytoskeletal perturbation. Biorheology. 2000;37(4):279–90.
7. Sung KL, et al. Dynamic changes in viscoelastic properties in cytotoxic T-lymphocyte-mediated killing. J Cell Sci. 1988;91(Pt 2):179–89.
8. Zhang H, et al. Myosin-X provides a motor-based link between integrins and the cytoskeleton. Nat Cell Biol. 2004;6(6):523–31.
9. Zeng Z, et al. Biophysical studies on the differentiation of human CD14 + monocytes into dendritic cells. Cell Biochem Biophys. 2006;45(1):19–30.
10. Faure S, et al. ERM proteins regulate cytoskeleton relaxation promoting T cell-APC conjugation. Nat Immunol. 2004;5(3):272–9.
11. Jiang Y, et al. Adhesion of monocyte-derived dendritic cells to human umbilical vein endothelial cells in flow field decreases upon maturation. Clin Hemorheol Microcirc. 2005;32 (4):261–8.
12. Steinman RM. Decisions about dendritic cells: past, present, and future. Annu Rev Immunol. 2012;30:1–22.
13. Gerszten RE, et al. MCP-1 and IL-8 trigger firm adhesion of monocytes to vascular endothelium under flow conditions. Nature. 1999;398(6729):718–23.
14. Yusuf-Makagiansar H, et al. Inhibition of LFA-1/ICAM-1 and VLA-4/VCAM-1 as a therapeutic approach to inflammation and autoimmune diseases. Med Res Rev. 2002;22 (2):146–67.
15. Nishibori M, Takahashi HK, Mori S. The regulation of ICAM-1 and LFA-1 interaction by autacoids and statins: a novel strategy for controlling inflammation and immune responses. J Pharmacol Sci. 2003;92(1):7–12.
16. van Wetering S, et al. VCAM-1-mediated Rac signaling controls endothelial cell-cell contacts and leukocyte transmigration. Am J Physiol Cell Physiol. 2003;285(2):C343–52.
17. Furie B, Furie BC, Flaumenhaft R. A journey with platelet P-selectin: the molecular basis of granule secretion, signalling and cell adhesion. Thromb Haemost. 2001;86(1):214–21.
18. Imhof BA, Aurrand-Lions M. Adhesion mechanisms regulating the migration of monocytes. Nat Rev Immunol. 2004;4(6):432–44.
19. Nguyen VA, et al. Adhesion of dendritic cells derived from CD34 + progenitors to resting human dermal microvascular endothelial cells is down-regulated upon maturation and partially depends on CD11a-CD18, CD11b-CD18 and CD36. Eur J Immunol. 2002;32(12):3638–50.
20. Burns S, et al. Maturation of DC is associated with changes in motile characteristics and adherence. Cell Motil Cytoskeleton. 2004;57(2):118–32.
21. Yao W, et al. Wild type p53 gene causes reorganization of cytoskeleton and therefore, the impaired deformability and difficult migration of murine erythroleukemia cells. Cell Motil Cytoskeleton. 2003;56(1):1–12.

22. Slivinsky GG, et al. Cellular electrophoretic mobility data: a first approach to a database. Electrophoresis. 1997;18(7):1109–19.
23. Wen Z, et al. Influence of neuraminidase on the characteristics of microrheology of red blood cells. Clin Hemorheol Microcirc. 2000;23(1):51–7.
24. Mehrishi JN, Bauer J. Electrophoresis of cells and the biological relevance of surface charge. Electrophoresis. 2002;23(13):1984–94.
25. Pollard TD, Borisy GG. Cellular motility driven by assembly and disassembly of actin filaments. Cell. 2003;112(4):453–65.
26. Pollard TD. Cellular motility powered by actin filament assembly and disassembly. Harvey Lect. 2002;98:1–17.
27. Heraud P, et al. Fourier transform infrared microspectroscopy identifies early lineage commitment in differentiating human embryonic stem cells. Stem Cell Res. 2010;4(2):140–7.
28. Naumann A, et al. Fourier transform infrared microscopy and imaging: detection of fungi in wood. Fungal Genet Biol. 2005;42(10):829–35.
29. Le Gal JM, et al. Conformational changes in membrane proteins of multidrug-resistant K562 and primary rat hepatocyte cultures as studied by Fourier transform infrared spectroscopy. Anticancer Res. 1994;14(4A):1541–8.
30. Le Gal JM, Morjani H, Manfait M. Ultrastructural appraisal of the multidrug resistance in K562 and LR73 cell lines from Fourier transform infrared spectroscopy. Cancer Res. 1993;53 (16):3681–6.
31. Steinman RM, Idoyaga J. Features of the dendritic cell lineage. Immunol Rev. 2010;234(1):5–17.
32. Steinman RM, Banchereau J. Taking dendritic cells into medicine. Nature. 2007;449 (7161):419–26.
33. Paul WE. Dendritic cells bask in the limelight. Cell. 2007;130(6):967–70.

Chapter 4
Biophysical Characteristics of DCs in Tumor Microenvironment

Abstract The generation and development of cancers are accompanied with a marked suppression of human immune system. The dysfunction of DCs has been implicated in tumor-bearing host. In order to elucidate the effects of tumor microenvironment-derived factors (TMDF) on the functions of DCs from biophysical aspects, DCs were treated with the cancer cells and their culture supernatant. The results showed that the biophysical characteristics of imDCs and mDCs were severely deteriorated by TMDF compared with those under normal conditions, moreover, and these changes are closely correlated with the expression levels of some cytoskeleton-binding proteins. The impaired biophysical characteristics of DCs may be one of many aspects of the immune escape mechanisms of tumors. It is significant for further understanding of the immune function of DCs, immune escape mechanism, and improving the clinical effectiveness of DCs-based immune therapy against cancer.

Keywords DCs · Biophysical characteristics · Tumor microenvironment-derived factors · Immune escape · Immune response

The DCs-based immune therapy is considered one of the most promising treatments to overcome cancers and has become a global hot research field, but there are many challenges in the clinical application of ex vivo DCs immunotherapy against cancer [1–4]. DCs are specialized antigen-presenting cells, which play a fundamental role in initiating and amplifying both the innate and adaptive immune responses [5–8]. The imDCs' precursors, present in non-lymphoid tissues, are efficient in antigen capture and subsequently travel via blood or lymph to lymphoid organs. During antigen capture and processing, mDCs express large amounts of peptide–MHC complexes and accessory molecules on their surfaces that are necessary for naive T cell activation [6, 8, 9]. Cytokines released by pathogens, e.g., lipopolysaccharide (LPS), and those locally produced, e.g., TNF-α, IL-1, are also mediators of DCs maturation and trigger migration of DCs towards the T cell areas of lymphoid organs [10]. DCs also can present the specific tumor antigens to T lymphocytes and elicit the effective antitumor immune responses [11]. DCs can conduct all elements of the immune

Z. Zeng et al., *Dendritic Cells: Biophysics, Tumor Microenvironment and Chinese Traditional Medicine*, SpringerBriefs in Biochemistry and Molecular Biology, DOI 10.1007/978-94-017-7405-5_4

orchestra, and they are therefore a fundamental target and tool for vaccination [2]. The tumor microenvironment, which is composed of immune cells, tumor cells, stromal cells, and the extracellular matrix, is a main battleground during the neoplastic process, involving proliferation, survival, and migration of tumor cells [4]. It has been generally acknowledged that there are many cytokines in the microenvironment of tumor tissues, namely, TMDF, such as transformed growth factors-β_1 (TGF-β_1), VEGF (vascular endothelial growth factors), IL-10 [12–15], gangliosides [16–18], and PGE$_2$ [18], etc., that solely or cooperatively suppress the antigen-presenting function of DCs, to result in immunotolerence of the tumor cells [19–22]. McBride et al. [23] suggested that IL-10 alters DC's functions via modulation of cell surface molecules to result in impaired T cell responses. Some researchers found that DCs were present in the tissues infiltrated by tumor cells, moreover, the quantities of DCs were closely related to the extent of tumor infiltration [16, 24–26]. While there have been some promising achievements, many challenges still remain in the clinical application of DCs-based therapeutic vaccination against cancer [1–4]. Although dysfunction of DCs in malignancy is believed to occur, the underlying mechanisms are poorly defined and appear to be heterogeneous in different types of cancers [27–31]. The physical organization of the plasma membrane of APC is important for both the regulation of antigen presentation and the induction of T cell proliferation, although different pathways may be involved in these two processes [32]. In order to elucidate immune escape mechanisms of tumors and to improve clinical efficacy of DCs-based therapy against cancer, the details about antigen presentation processes of DCs under normal and pathological conditions need to be explored from different aspects. The interactions between DCs and T cells in lymph node have been investigated by various groups with different methods [33–37]. It was found that different physical contact modes cause different results of immune responses. From our perspective, biophysical characteristics of DCs may play a key role in the initiation of immune responses. Our previous studies suggested that DCs at different differentiation stages manifest various biophysical behaviors [38, 39]. Here, DCs including imDCs and mDCs were treated with supernatant of Jurkat cells (SJC) and cancer cells including hepatocellular carcinoma cells (HCCs) and leukemia cells (K562) in order to clarify the immune escape mechanisms of tumors by exploring the relationship between the tumor microenvironment and the biophysical characteristics of DCs.

Cell viability could affect the biophysical characteristics and immune functions of DCs. The results (Table 4.1) showed that the viabilities of DCs were not affected by SJC.

The CD1a, CD40, CD80, CD83, CD86, and HLA-DR are important signaling and costimulatory molecules involved in the antigen presentations. As shown in Table 4.2; Figs. 4.1 and 4.2, the expression levels of these surface marker molecules on DCs were downregulated by TMDF, indicating that the immune phenotypes of DCs were impaired by TMDF. This result is consistent with other reports [4, 40, 41].

The fluid state of the lipid bilayer is thought to play an important role in cellular function and processes, such as growth and differentiation [42]. Each of these processes may be accompanied by specific alterations in membrane fluidity, which is

Table 4.1 Viabilities of cells in different culturing media assessed by trypan blue staining[a]

	MOs	imDCs	imDCs + SJCs	mDCs	mDCs + SJCs
Living cells (%)	97.7 ± 1.1	93.5 ± 1.2[*]	92.4 ± 1.3	91.3 ± 1.4[*]	90.2 ± 1.3

Comparison of imDCs and mDCs with imDCs + SJC and mDCs + SJC: [*]$p < 0.05$
[a]All values are expressed as mean ± SD (cited from Cell Motility and Cytoskeleton, 2007, 64 (3):186–198)

Table 4.2 The effects of SJC on the phenotypes of DCs

Group	Positive expression rate						
	CD1a	CD11c	CD40	CD80	CD83	CD86	HLA-DR
Control							
imDCs	63.7 ± 2.1	47.8 ± 3.2	39.1 ± 4.2	23.5 ± 1.7	14.3 ± 1.6	58.2 ± 3.5	71.8 ± 2.6
mDCs	60.2 ± 1.8	86.3 ± 2.8	87.3 ± 3.6	92.3 ± 2.2	90.5 ± 1.9	96.8 ± 2.9	98.3 ± 3.2
Treatment							
imDCs + SJC	54.8 ± 1.6	45.2 ± 2.6	31.6 ± 2.3[*]	18.6 ± 1.2[*]	13.9 ± 2.2	41.5 ± 1.6[*]	55.7 ± 3.4[*]
mDCs + SJC	51.5 ± 1.9	84.7 ± 3.1	62.9 ± 3.7[*]	83.5 ± 1.5[*]	71.3 ± 2.9[*]	85.2 ± 3.1[*]	86.7 ± 1.7[*]

Comparison of DCs + SJC with corresponding DCs: [*]$p < 0.01$. All values are expressed as mean ± SD (cited from Cell Motility and Cytoskeleton, 2007, 64(3):186–198)

associated with changes in composition of membrane lipids such as cholesterol/phospholipid levels and the degree of fatty acyl chain saturation within the cell membrane [43]. The fluorescence polarization parameter p (Figs. 4.3 and 4.4) markedly increased in the DCs treated with TMDF, suggesting that the microfluidity of membrane lipid molecules of DCs was decreased by TMDF. Changes in membrane fluidity may be associated with variations in the exposure of surface receptors or ligands and antigen-peptides [44]. For imDCs, impaired membrane fluidity might contribute to the decrease in cell deformability and thus impair their proper motilities directed by chemokines or their receptors, leading to a diminution of the number of imDCs that would reach the tumor tissue. This is in concert with the notion that the quantities of DCs are negatively correlated with the extent of tumor infiltration [27, 29–31]. For mDCs, the decreased fluidity of membrane lipid molecules by the tumor microenvironment could affect their inter-actions with naive T lymphocytes and disturb the formation of immune synapse, thus suppressing the activation of the latter. Our results support the viewpoint that altered signals mediated by DCs–T cell contact are the primary mechanisms for impaired stimulatory capacity of DCs in tumor microenvironments in vivo [32].

The osmotic fragility of the cell reflects its ability to resist the hypoosmolality-induced lysis. As showed in Fig. 4.5, the osmotic fragilities of DCs were increased by the TMDF, indicating an impaired ability of DCs to withstand

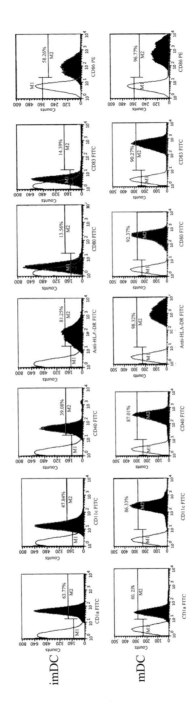

Fig. 4.1 The expression profiles of costimulatory molecules in imDCs and mDCs. The *first panels* are for imDCs and the *second panels* are for mDCs. The *empty curves* are for negative controls (cited from Cell Biochemistry and Biophysics, 2009, 55(1):33–43)

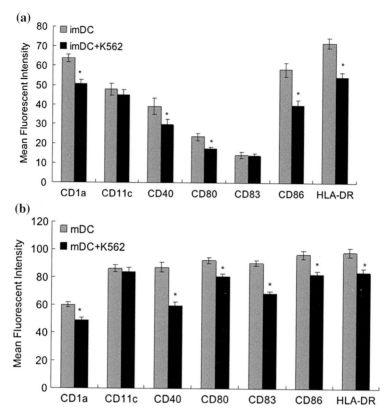

Fig. 4.2 Phenotypic marker analysis and apoptosis analysis by flow cytometry on DCs and K562-conditioned DCs. **a** The expression of phenotypic markers in imDCs and K562-conditioned imDCs (imDCs + K562). **b** The expression of phenotypic markers in mDCs and K562-conditioned mDCs (mDCs + K562). Compared with corresponding DCs: $^*p < 0.05$ (cited from Journal of Biomechanics, 2010, 43(12):2339–2347)

the decreased osmotic pressure. Tumors can damage or disrupt their surrounding tissue or trigger a stress response when their oxygen and nutrient demands outstrip their supplies. These processes can cause pH imbalance because of the metabolic disturbance and the generation of reactive oxygen species, thus leading to alterations in local osmotic pressure and the lyses of DCs with increased osmotic fragility. This might be another reason for the poor immunogenesis of tumor.

The micropipette experiment evaluates the viscoelastic properties of cells. The results (Tables 4.3 and 4.4) of this study showed that DCs treated with HCCs and K562 had significantly higher viscoelasticity or less deformability. This could be the reason of impaired motilities of DCs in cancer microenvironment.

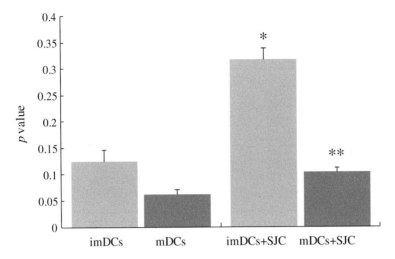

Fig. 4.3 Fluorescence polarization of DCs under normal condition and following treatment with the tumor-conditioned medium, i.e., supernatant of Jurket cells (SJC). The data represent mean ± SD. Compared with corresponding untreated DCs, the membrane lipid fluidity (inversely related to the p value of fluorescence polarization) of DCs + SJC significantly decreased ($^*p < 0.01$ and $^{**}p < 0.05$, receptively). These measurements were performed in triplicates under the same condition (cited from Cell Motility and Cytoskeleton, 2007, 64(3):186–198)

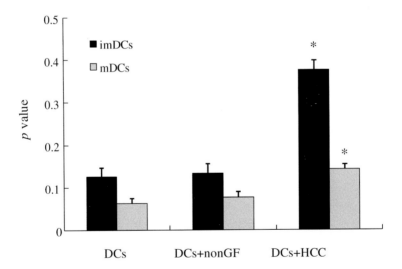

Fig. 4.4 The fluorescence polarization parameter (p) of DCs. DCs cultured with and without growth factor (GF) and co-cultured with HCCs are labeled as DCs, DCs + nonGF, and DCs + HCC, respectively. Compared with DCs, $^*p < 0.05$ (cited from Cell Biochemistry and Biophysics, 2009, 55(1):33–43)

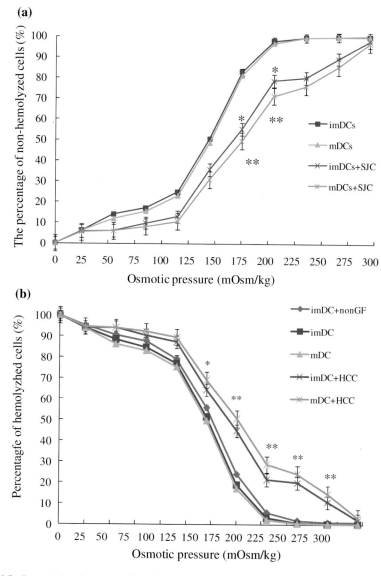

Fig. 4.5 Osmotic fragility curves for DCs under different culturing conditions. **a** DCs treated with SJC. **b** DCs treated with HCC. The data represent mean ± SD. These measurements were performed in triplicates under the same condition (cited from Cell Motility and Cytoskeleton, 2007, 64(3):186–198) and Cell Biochemistry and Biophysics, 2009, 55(1):33–43)

The electrical characteristics of cell plasma membrane play important roles in its interaction with cells and extracellular matrix. In vivo, both the surfaces of DCs and T cells covered with negatively charged glycocalyx components. During the process of antigen presentation, DCs and T cells undergo a direct physical contact in

Table 4.3 The viscoelastic coefficients of DCs under different conditioned media (mean ± SD)

$n = 20$	+nonGF		+HCC		DCs	
	imDCs	mDCs	imDCs	mDCs	imDCs	mDCs
K_1 (N/m²)	32.65 ± 4.66	20.22 ± 3.28	43.74 ± 5.22*	29.08 ± 3.27*	21.62 ± 3.61	15.78 ± 2.93
K_2 (N/m²)	16.87 ± 3.39	12.76 ± 2.41	23.75 ± 2.64*	19.77 ± 2.80*	11.31 ± 4.72	8.33 ± 1.52
μ (N S/m²)	11.29 ± 2.11	8.36 ± 1.24	10.56 ± 2.87*	7.38 ± 2.21*	4.53 ± 1.51	4.16 ± 1.58

Comparison of DCs + HCC with corresponding DCs and DCs + nonGF: *$p < 0.05$ (cited from Cell Biochemistry and Biophysics, 2009, 55(1):33–43)

Table 4.4 The viscoelastic coefficients of un-conditioned and K562-conditioned DCs (x ± SD)

$N = 20$	DCs		K562-conditioned DCs	
	imDCs	mDCs	imDCs	mDCs
K_1 (N/m²)	21.62 ± 3.61	15.78 ± 2.93	42.75 ± 3.67*	27.36 ± 2.17*
K_2 (N/m²)	11.31 ± 4.72	8.33 ± 1.52	24.96 ± 2.64*	18.92 ± 1.83*
μ (NS/m²)	4.53 ± 1.51	4.16 ± 1.58	10.30 ± 1.82*	8.31 ± 1.16*

Compared with corresponding un-conditioned DC: *$P < 0.05$ (cited from Journal of Biomechanics, 2010, 43(12):2339–2347)

which they must overcome the barrier posed by glycocalyx. The lower surface charges of DCs treated with TMDF (Fig. 4.6) could decrease the repulsion forces and enhance their encountering probabilities with other cells. Previous findings [33–35] indicate that, during the interactions between DCs and T cells, the different contact times induce different immune responses. Considering the notion that the longer contact times between DCs and T cells prime immune tolerance [34], the decreased surface charges on mDCs under cancer-conditioned culture might enhance the contact times of DCs and T cells and affect their interaction processes leading to antigen-specific immune tolerance in the host loading tumor.

The biophysical properties of cells, including viscoelasticity, osmotic fragility etc., are determined by the organization of the cytoskeleton. Any change in these parameters would affect the deformation and the migration of the cells. These properties may be related to their cytoskeletons and influence their migration. The actin cytoskeleton system of immune cells plays a crucial role in the regulation of their motility and signaling [45, 46]. Confocal laser scanning microscopy was used to study the cytoskeleton organizations in DCs under different conditioned media. The F-actin of imDCs and mDCs was almost entirely (∼85 %) distributed under the plasma membrane of the cell and rarely in cytoplasm (Figs. 4.7, 4.8a and 4.9a), and these results are in agreement with those reported by others [47]. In contrast, the F-actin organizations in DCs treated with tumor-derived factor were disorganized and found mostly in the cytoplasm (75–80 %). The disordered organizations of F-actin in DCs under cancer-conditioned culture might affect the migration capabilities of these cells, which also have poor deformability and increased

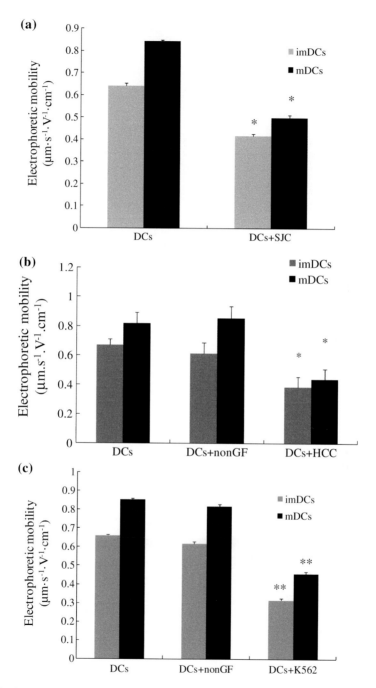

Fig. 4.6 The electrophoretic mobility of DCs. **a** SJC; **b** HCC (cited from Cell Biochemistry and Biophysics, 2009, 55(1):33–43); **c** K562. Compared with DCs, $^*p < 0.05$ or $^{**}p < 0.01$

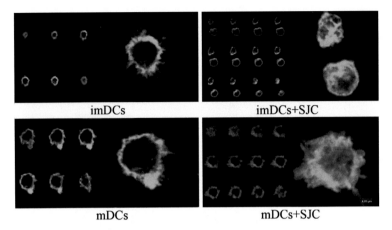

Fig. 4.7 Representative series of CLSM images (*left*) and three dimensional images (*right*) of rhodamine phalloidin-labeled F-actin DCs. The serial sectional images (*left half* of each of the four groups) were reconstructed into the three-dimensional image (*right half* of each group) (cited from Cell Motility and Cytoskeleton, 2007, 64(3):186–198)

osmotic fragility. Most leukocyte subsets have common traits regarding cytoskeletal reorganization [48]. The cytoskeleton is the structural basis of cell morphology, cell motility, DCs–naive T cells interactions, and signal transduction. It could be inferred that the disordered cytoskeleton of cells in tumor microenvironment would not only impair the motility or migration of imDCs, but also hamper the interaction processes (the formations of immune synapses) between mDCs and naive T cells in vivo. Our results also showed that the F-actin expression levels of DCs treated with cancer-derived factors were significantly higher than those of untreated imDCs and mDCs (Table 4.5; Figs. 4.8b and 4.9b). It could be inferred that the extension processes of F-actin in DCs appeared abnormal. The characteristics of disordered organization and higher expression levels of actin cytoskeleton as well as the decreased membrane fluidity of DCs treated with TMDF could contribute to the poor deformability and impaired motility of DCs.

An important objective in the DCs-based immunotherapy is that the DCs generated in vitro should maintain their motile capacity such that they can migrate from the site of injection to T cell zones of secondary lymphoid organs to present the acquired antigens for the activation of specific T cells. Therefore, transendothelial migration measurements were performed with the goal of assessing the effect of TMDF on DCs motility. The results of transendothelial migration assay (Fig. 4.10) demonstrated that the migration capabilities of DCs indeed were impaired by TMDF. There are dysfunctional expressions of chemokines (e.g., CCR7) in DCs under tumor microenvironments [49, 50], which could lead to their improper

(a)

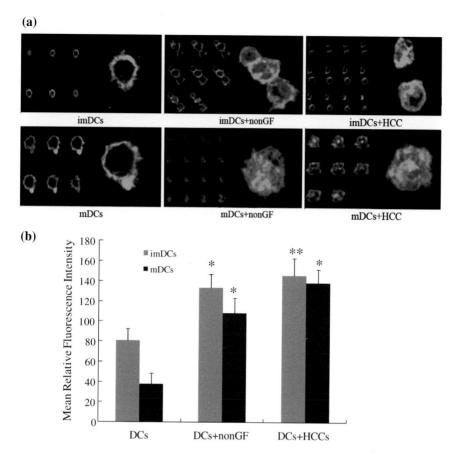

(b)

Fig. 4.8 **a** The serial sectional images (*left panel* of each image) were reconstructed into the three-dimensional image (*right panel* of each image). **b** The expression levels of F-actin in DCs were quantified by measuring the mean relative fluorescent intensities of the images. Compared with corresponding DCs: $^*p < 0.05$ or $^{**}p < 0.01$ (cited from Cell Biochemistry and Biophysics, 2009, 55(1):33–43)

migration in vivo. Cells with high viscoelasticity or poor deformability cannot form strong adhesion with other cells or tissues because of their inability to increase contact areas. Hence, the TMDF-conditioned DCs would have greater difficulty to attach onto the vascular endothelium or extracellular matrix, especially in the presence of the shear stress due to blood flow, and to subsequently emigrate from the site of antigen uptake to secondary lymphoid tissues. To perform their

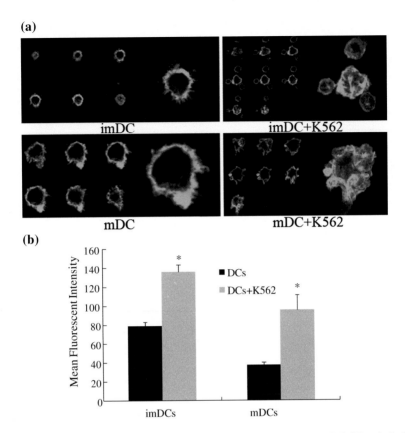

Fig. 4.9 a Representative series of CLSM images of rhodamine phalloidin-labeled F-actin in DCs (including imDC and mDC) and K562-conditioned DCs (imDCs + K562 and mDCs + K562). The serial sectional images were reconstructed into the three-dimensional images (*right panels*). **b** Mean fluorescent intensity of F-actin signals. Compared with corresponding DCs: $^*p < 0.05$ (cited from Journal of Biomechanics, 2010, 43(12):2339–2347)

Table 4.5 The mean relative fluorescence intensities (*I*) of DCs and DCs + SJC[a]

$N = 20$	imDCs	mDCs	imDCs + SJC	mDCs + SJC
I	85.22 ± 2.83	51.76 ± 3.25	134.94 ± 5.23	107.56 ± 3.51

Comparison of imDCs + SJC with imDCs: $^*p < 0.001$; comparison of mDCs + SJC with mDCs: $^{**}p < 0.005$
[a]All values are expressed as mean ± SD (cited from Cell Motility and Cytoskeleton, 2007, 64 (3):186–198)

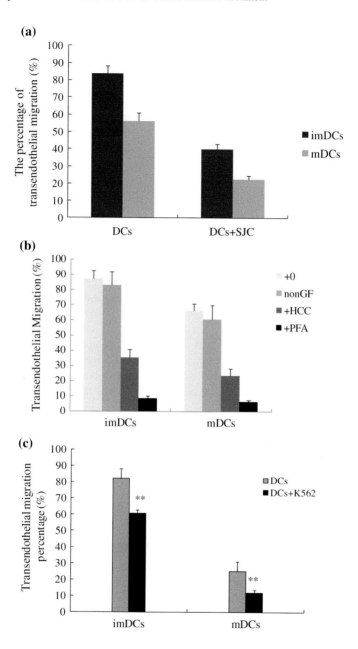

◄ **Fig. 4.10** Transendothelial migration percentages of DCs. DCs (10^6) or DCs treated with TMDF were added onto the HUVEC monolayers and incubated at 37 °C for 12 h. Cells collected from the lower compartment face of the membrane and in the solution of the lower compartment were counted with a FACScan (Becton Dickinson). The ratio of the sum of cell count collected below the membrane to that added to the chamber yields the transendothelial migration percentage of cells. The data represent mean ± SD. These measurements were performed in triplicates under the same condition. **a** SJC (cited from Cell Motility and Cytoskeleton, 2007, 64(3):186–198); **b** HCC (cited from Cell Biochemistry and Biophysics, 2009, 55(1):33–43); **c** K562 (cited from Journal of Biomechanics, 2010, 43(12):2339–2347). Compared with corresponding DCs: $^*p < 0.05$ and $^{**}p < 0.01$

physiological functions, DCs must possess appropriate motility. For DCs-based ex vivo immune therapy against tumors, the DCs loaded with tumor antigens also must maintain excellent motility to migrate from the injection location to secondary lymphoid tissues and initiate tumor-specific immune response. Thus, it could be explained why the overall number of DCs reaching a lymph node is very small (probably less than 1 %) after their intracutaneous injection into a host loading tumor [2, 51] and support the notion that the biophysical properties of the cell and/or its environment can influence the complex signaling events that direct DCs trafficking towards lymphatic vessels or lymph node in vivo [52].

The endocytic activity and stimulatory capacity of DCs are their most important functions. The endocytic and mixed lymphocyte reaction (MLR) assays were performed to further establish the effects of TMDF on the endocytic activities of imDCs and stimulatory capacities of mDCs and to assess the implications of our biophysical measurements. The results (Figs. 4.11 and 4.12) indicated that the endocytic function and the stimulatory capacities to activate T cell proliferation of DCs were markedly suppressed by TMDF. To further demonstrate the crucial role of biophysical characteristics of DCs in their immune functions, imDCs and mDCs were treated by 1 % paraformaldehyde (PFA) prior to the endocytic and MLR assays. The low concentration of paraformaldehyde can affect the biophysical characteristics of cells, but not the expressions of their surface marker molecules [32]. Our results showed that the PFA-treated DCs lost their stimulatory and endocytic potentials almost totally. While the downregulated surface marker molecule expressions of mDCs by the TMDF (Table 4.2; Figs. 4.1 and 4.2) can affect their costimulatory potentials, the mDCs and mDCs fixed with PFA almost lost their costimulatory potentials totally (Figs. 4.11 and 4.12). This finding provides direct support for the notion that biophysical characteristics of DCs could be related to the regulation of antigen presentation and activations of T cell processes via different pathways. Therefore, the biophysical characteristics and molecular expressions of DCs are both critically important for antigen presentation and activations of T cells.

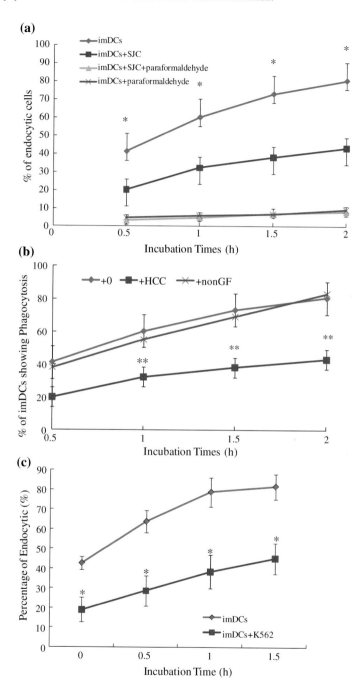

◀ **Fig. 4.11** The Endocytic activities of imDCs. Cells (2×10^5) were incubated at 37 °C with 25 μg/mL FITC-dextran particles (43.2 kDa). The numbers of cells that had endocytosed the particles were determined by flow cytometry. The data represent mean ± SD. These measurements were performed in triplicates under the same condition. **a** SJC (cited from Cell Motility and Cytoskeleton, 2007, 64(3):186–198); **b** HCC (cited from Cell Biochemistry and Biophysics, 2009, 55(1):33–43); **c** K562 (cited from Journal of Biomechanics, 2010, 43(12):2339–2347). Compared with corresponding DCs: $^*p < 0.05$ and $^{**}p < 0.01$

Vibrational spectroscopy offers complete information on the chemical composition of samples, regarding both major and minor compounds, which presents many characteristic bands in the infrared range (IR). The ratio of A_{1121}/A_{1545} corresponds to an RNA/amide, which reflects the gene transcription states of cells. As shown in Table 4.5, the ratios of A_{1121}/A_{1545} of DCs + HCC were significantly lower than those of DCs ($^*P < 0.05$) and there were no difference among DCs + HUVEC, DCs + HC and DCs ($P > 0.05$), suggesting that the gene transcription states of DCs were specifically suppressed by HCC, and this is consistent with others [53–55]. The inhibition of gene transcription state might be associated with the abnormal expressions of some important proteins in DCs leading to the dysfunctional immune responses of the host. Under the tumor microenvironment, NF-κB plays an important role in cancer development and progression because it is a crucial transcription factor which regulates immune functions [56]. Kuwabara and colleagues found that the NF-κB signaling pathways perform a critical role in CCR7-mediated IL-23 production [57]. The RelB subunit of NF-κB controls DC's maturation and may be therapeutically targeted to manipulate the T cell response in disease. Several groups have reported that RelB promoted DC activation, and RelB-silenced DCs could induce donor-specific hyporesponsiveness and impair immune functions of T cells [58, 59]. The TMDF downregulate the expression levels of chemokines, adhesion molecule, and costimulatory molecules by suppressing expression of NF-κB [4, 60]. Therefore, the expression levels of NF-κB in cells were selected as one of the indicators of DC's functional states. To confirm the results of FTIR, DCs were treated with NF-κB inhibitor ASA, the data (Table 4.6) showed that the ratio of A_{1121}/A_{1545} of DCs was indeed markedly decreased by ASA, and this was consistent with the data of western blotting (Fig. 4.13). Moreover, the ratio of A_{1121}/A_{1545} (Fig. 4.14) was closely correlated with the expression levels of NF-κB (R^2: 0.69 and 0.81). Thus, these results support the idea that FTIR could be clinically applied to estimate the functional states of DCs.

The ratios of A_{1030}/A_{1080} and A_{1030}/A_{2924}, respectively, correspond to glucose/phospholipids and glucose/phosphate, which reflect the energy states of cells. As shown in Table 4.1, the ratios of A_{1030}/A_{1080} and A_{1030}/A_{2924} of DCs + HCC were lower than those of DCs ($P < 0.05$) and there was no difference

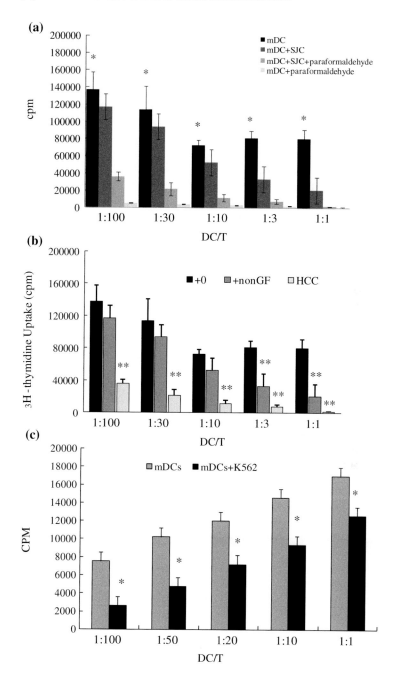

◄ **Fig. 4.12** Stimulatory abilities of mDCs in the MLR assay under different culture conditions. mDCs, used as stimulators, were generated from fresh PBMC progenitors. Allogeneic T cells, served as responders, were obtained from PBMC of other people. mDCs were added in graded doses (10^3–10^5 cells/well) to T cells (1×10^5), and proliferation of T cells was measured by the uptake of ^{3}H-thymidine. The data represent mean ± SD. These measurements were performed in triplicates under the same condition. **a** SJC (cited from Cell Motility and Cytoskeleton, 2007, 64 (3):186–198); **b** HCC (cited from Cell Biochemistry and Biophysics, 2009, 55(1):33–43); **c** K562 (cited from Journal of Biomechanics, 2010, 43(12):2339–2347). Compared with corresponding DCs: $^*p < 0.05$ and $^{**}p < 0.01$. The imDCs treated by 1 % PFA lost their stimulatory activities almost totally

among DCs + HUVEC, DCs + HC, and DCs ($P > 0.05$), indicating that the energy states of DCs were also significantly and specifically impaired by HCC. It could be inferred that the insufficient glucose content of DCs could not be able to afford enough energy to perform their physiological functions, such as migration into peripheral tissue and interaction with naive T cells in lymph node. This could explain why the overall number of DCs reaching a lymph node is very small (probably less than 1 %) after their intracutaneous injection into a host loading tumor [61, 62]. In addition, the insufficient energy supplies caused by tumor cells-derived factors might affect the other functions of DCs, such as the metabolisms of protein and lipid, this inevitably might be associated with the uptake, processing, and presenting of antigen of DCs. It was one of the possible reasons of the deteriorated motilities and immune regulatory functions of DCs under tumor microenvironment. Taken together, the present study suggested that gene transcriptional activity and energy states of DCs were inhibited by HCCs, this might be one of aspects of tumor immune escape mechanisms. The ratios of absorption intensity of FTIR at given wave number were closely correlated with the expression levels of NF-κB. It laid the foundation for the application of FTIR to the identification of functional states in the DCBV preparation protocol.

An important action of IL-12 is its induction of other cytokines, particularly INF-γ and IL-18, which coordinate the ensuing immune responses. As shown in Fig. 4.15, IL-12 productions of both imDCs and mDCs were significantly inhibited by SJC, suggesting that the immune regulatory functions of these DCs were also impaired by the factors derived from the tumor. This finding is in agreement with those from other reports [41, 63].

IL-10, VEGF, and TGF-β$_1$ were selected for measurement because of their potent suppressive actions. It was found that IL-10, VEGF, and TGF-β$_1$ were all highly excreted by the Jurkat cells (Fig. 4.16). These suppressive cytokines could be the major contributors to impair biophysical functions and immune capability of DCs in the presence of SJC.

Table 4.6 The ratios of absorption intensity at given wave number in DCs under different conditioned microenvironments ($X \pm SD$)

Types of DCs		A_{1020}/A_{1545} DNA/amide II	A_{1121}/A_{1545} RNA/amide II	A_{1030}/A_{1080} glucose/phospholipid	A_{1030}/A_{2924} glucose/phosphate
DCs	imDCs	3.782 ± 0.016	2.953 ± 0.060	1.274 ± 0.046	0.753 ± 0.033
	mDCs	0.531 ± 0.032	1.219 ± 0.039	0.169 ± 0.022	0.957 ± 0.058
DCs + nonGF	imDCs + nonGF	3.655 ± 0.027	1.822 ± 0.086	1.383 ± 0.031	0.764 ± 0.028
	mDCs + nonGF	0.583 ± 0.026	0.877 ± 0.051	0.423 ± 0.015	0.604 ± 0.022
DCs + HUVEC	imDCs + HUVEC	3.452 ± 0.187	2.769 ± 0.171	1.302 ± 0.054	0.748 ± 0.048
	mDCs + HUVEC	0.549 ± 0.079	1.233 ± 0.089	0.152 ± 0.103	0.933 ± 0.021
DCs + HC	imDCs + HC	3.556 ± 0.045	2.843 ± 0.062	1.266 ± 0.082	0.712 ± 0.067
	mDCs + HC	0.586 ± 0.034	1.176 ± 0.095	0.149 ± 0.061	0.926 ± 0.043
DCs + HCC	imDCs + HCC	3.721 ± 0.027	$0.603 \pm 0.004^{*}$	$0.382 \pm 0.010^{*}$	$0.302 \pm 0.017^{*}$
	mDCs + HCC	0.577 ± 0.027	$0.361 \pm 0.021^{*}$	$0.423 \pm 0.007^{*}$	$0.408 \pm 0.023^{*}$
DCs + ASA	imDCs + ASA	3.718 ± 0.018	$0.296 \pm 0.021^{**}$	$0.351 \pm 0.012^{**}$	$0.281 \pm 0.024^{**}$
	mDCs + ASA	0.488 ± 0.024	$0.125 \pm 0.009^{**}$	$0.107 \pm 0.009^{**}$	$0.386 \pm 0.031^{**}$

imDCs + HCC and mDCs + HCC respectively compared with imDCs and mDCs: * $p < 0.05$; imDCs + ASA and mDCs + ASA respectively compared with imDCs and mDCs: ** $p < 0.01$ (cited from Biomedical Engineering Online. 2014, 13(1):2)

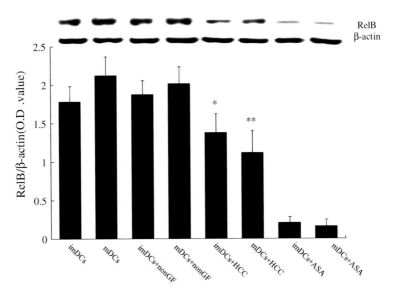

Fig. 4.13 The expression levels of RelB in DCs under different conditioned microenvironments. Cells were lysed with RIPA buffer (20 mM sodium phosphate, pH 7.4, 150 mM sodium chloride, 1 % Triton X-100, 5 mM EDTA, 200 μM phenymethylsulfonyl fluoride, 1 μg/ml aprotinin, 5 μg/ml leupeptin, 1 μg/ml pepstatin, and 500 μM Na_3VO_4). The protein extracts were electrophoresed on 12–14 % SDS-polyacrylamide gel and transferred onto a nitrocellulose membrane (Invitrogen, USA). After blocking with 5 % BSA in 0.1 % Tween 20 in PBS, membranes were probed with primary antibodies. Anti-RelB and anti-β-actin antibodies (Sigma) were diluted in blocking buffer and incubated with the blots overnight at 4 °C. The bound primary antibodies were probed with a 1:2000 diluted secondary antibody (goat anti-human IgG-HRP antibody) and visualized by the ECL chemiluminescence system (Amersham, USA). The gray values of proteins were measured by Image J (1.45). The expression levels of proteins were normalized to those of corresponding β-actin. Compared with DCs: $^*p < 0.05$ or $^{**}p < 0.01$ (cited from Biomedical Engineering Online. 2014, 13(1):2)

Trypan Blue staining assay (Table 4.1) and apoptosis analyses (Fig. 4.17) showed that the TMDF did not affect the viability and apoptosis of DCs, suggesting that the TMDF caused the impairment of biophysical characteristics of DCs rather than cell death.

To explore the potential molecular mechanisms of above-mentioned data, the microarray analysis and two-dimensional electrophoresis-based proteomics were performed. Microarray analysis (Table 4.7) showed that the imDCs treated K562 had 1121 genes with ≥ 2-fold change ($p < 0.05$) compared with un-conditioned imDCs and that the mDCs treated with K562 had 988 genes with ≥ 2-fold change ($p < 0.05$) compared with untreated mDCs. Gene ontology analysis showed that these genes are associated with cellular functions such as immunology, metabolism, cell migration, apoptosis, etc. Some interesting genes were selected for confirmation (Fig. 4.18).

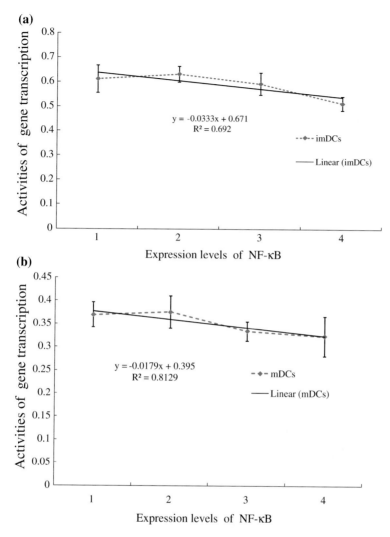

Fig. 4.14 Analyses of linear regression of the activities gene transcription and the expression levels of NF-κB. **a** imDCs (R^2: 0.69), **b** mDCs (R^2: 0.81) (cited from Biomedical Engineering Online. 2014, 13(1):2)

Two-dimensional gel electrophoresis was performed to identify the differentially expressed proteins between DCs and K562-treated DCs. As shown in Fig. 4.19, the overall patterns of un-conditioned and K562-conditioned DCs were generally similar, but several protein changes were reproducibly detected. Nineteen proteins of interest were excised from the gels and analyzed. Analysis of the spectra by the MS-FIT search program identified the differentially expressed proteins to be members of specific families, including cytoskeleton and its related proteins (actin,

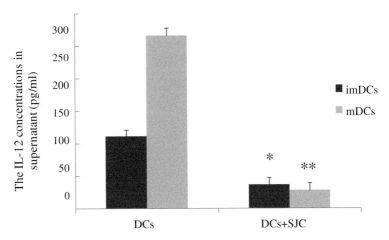

Fig. 4.15 IL-12p40 production of DCs. The concentration of IL-12p40 in the supernatants of DCs treated with SJC were measured by ELISA. The data represent mean ± SD. These measurements were performed in triplicates under the same condition. Compared with corresponding DCs: *$p < 0.05$ and **$p < 0.01$ (cited from Cell Motility and Cytoskeleton, 2007, 64(3):186–198)

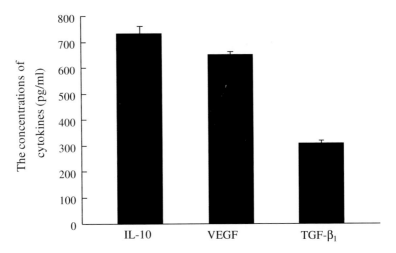

Fig. 4.16 Cytokine concentrations in the supernatant of Jurkat cells. The supernatant of Jurkat cells had high levels of IL-10, VEGF, and TGF-β_1 in comparison with those of nontumor cell lines. The data represent mean ± SD. These measurements were performed in triplicates under the same condition (cited from Cell Motility and Cytoskeleton, 2007, 64(3):186–198)

cofilin1, profilin1, stathmin1), Ca^{2+}-binding proteins, as well as signaling proteins (Table 4.8). The protein expressions of profolin1, cofilin1, and p-cofilin1 were validated by Western blot (Fig. 4.19b, c).

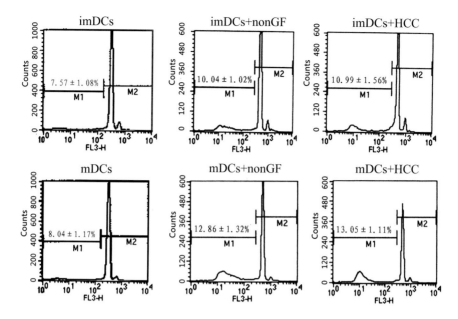

Fig. 4.17 The apoptosis analysis by flow cytometry on DCs. The cells in the low fluorescence area (M_1) are apoptotic cells. DCs had relatively few apoptotic cells, DCs + nonGF and DCs + HCCs had a higher proportion of cells undergoing apoptosis ($^*p < 0.05$), whereas there was no difference between DCs + nonGF and DCs + HCCs ($p > 0.1$) (cited from Cell Biochemistry and Biophysics, 2009, 55(1):33–43)

Table 4.7 The fold changes of differentially expressed genes under different culture conditions

Name	Genebank ID	K562-conditioned imDCs/imDCs	K562-conditioned mDCs/mDCs
TIMP1	NM_003254	6.4984	2.3287
NDRG1	NM_006096	6.7659	3.4707
HMG1	NM_002128	0.2918	0.5916
VEGFA	NM_002375	1.5509	1.8083
MMP9	NM_004994	0.4566	0.8300
IFNG	NM_000619	0.4730	0.3186
ASGR1	NM_001671	2.6924	1.5447
BNIP3	NM_004052	12.2191	16.9103
DUSP1	NM_004417	6.5972	1.5846
HIF1	NM_001530	0.5918	0.9204
HSPA6	NM_002155	15.7770	21.5965
IL6	NM_002155	0.4083	0.4377
IL-8	M17017	0.0886	0.4053
TGFβ$_1$	NM_000660	1.3272	1.4797

Cited from Journal of Biomechanics, 2010, 43(12):2339–2347

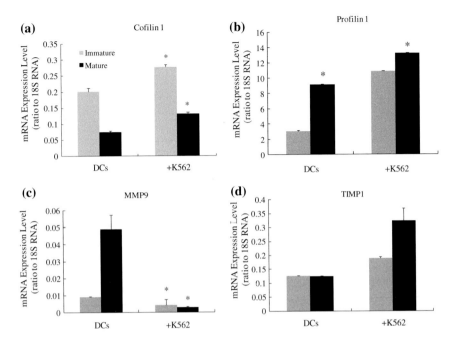

Fig. 4.18 Effects of K562 on mRNA expressions of cofilin 1 (**a**), profilin 1 (**b**), MMP9 (**c**), and TIMP1 (**d**) in DCs as determined by real-time PCR. Total RNA was isolated from DCs and K562-conditioned DCs (DCs + K562). Expression levels were calculated from cycle threshold (CT) values and normalized to that of 18 S RNA. Compared with corresponding DCs: $^*p < 0.05$ (cited from Journal of Biomechanics, 2010, 43(12):2339–2347)

The data showed that exposure to K562 has caused changes in the expression of several genes and proteins in DCs, some of which are listed in Tables 4.7 and 4.8. The identified genes and proteins are relevant to many cellular functions such as metabolism, signal transduction, viscoelasticity, deformability, motility, etc., and many of them are actin-binding proteins (ABP). As shown in Fig. 4.19, the expressions of cofilin1, p-cofilin1 (inactive form of cofilin1), and profilin1 all increased significantly in K562-conditioned DCs than those in un-conditioned DCs. The changes in these proteins and other ABPs, such as calponin, stathmin1, and Rho GDI proteins (Table 4.8), may lead to the cytoskeleton rearrangement and deformability/motility deteriorations found in K562-conditioned DCs [64–66], thus disturbing these structural and functional features that are critical for DCs to perform normal immune functions. In addition, the balance between matrix metallo-proteinases 9 (MMP9) and its inhibitor TIMP-1 contribute to the degradation or deposition of the extracellular matrix (ECM) [67, 68]. The changes in expression of MMP9 and TIMP1 genes in K562-conditioned DCs (Fig. 4.17) could result in the changes in their protein expression, which may be another reason for their defective immune functions.

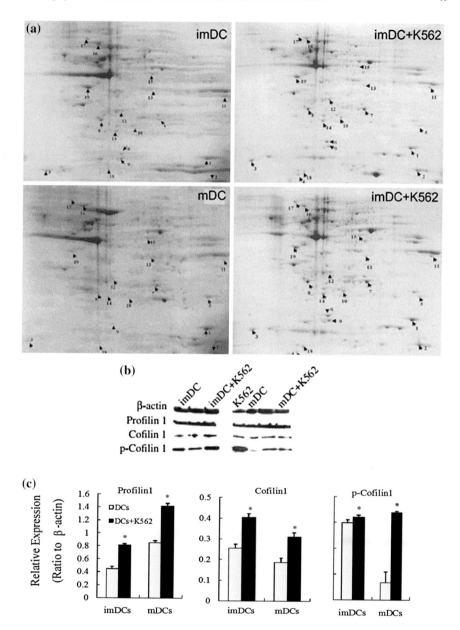

Fig. 4.19 a Two-dimensional gel profiles of DCs and K562-conditioned DCs (DCs + K562). *Arrows* denote the identified protein number (*left* PI3, *right* PI10). b Effect of K562 on expressions of profilin1, cofilin1, and p-cofilin1 assessed by Western blotting. β-Actin was used as a control. c Quantification data for Western blot (cited from Journal of Biomechanics, 2010, 43(12):2339–2347)

Table 4.8 Identification by mass spectrometry of the regulated protein spots when DCs were cultured under different conditions

Spot number	Protein name	Relative volume imDCs: K562-conditioned imDCs	Relative volume mDCs: K562-conditioned mDCs	Mr/pI Databank	Accession number SwissPort
1	Cofilin1	1:1.5	1:2	18.3/8.26	P23528
2	Profilin1	1:1.5	1:1.5	14.9/8.48	P07737
3	Calmodulin	1:3	1:2	16.7/4.09	P62158
4	Thioredoxin	_:2	_:1.5	11.6/4.8	P10599
5	Thioredoxin peroxidase	_:1.5	1:1	41.2/9.1	Q63716
6	Superoxide dismutase [Cu-Zn]	1:2	_:2	10.3/7.19	GI/223020
7	Tim	1:3	1:2	26.5/6.51	GI/999893
8	Rac rhoGDI complex	1:2	1:4	20.5/6.16	GI/9955206
9	Stathmin1	1:2	1:3	17.3/5.76	GI/15680064
10	PYD/CARD domain	1:1.5	1:2	21.6/5.95	GI/10835256
11	HnRNPA2/B1	_:1.5	1:3		GI/26662781
12	Pa28a	1:1	1.5:1	28.7/5.78	GI/54695544
13	Calponin 2 isoform 2	_:1	1:1	33.6/6.95	GI/4758018
14	Rho GDI 2	1:2	_:2	23.0/5.1	P52566
15	Actin	1:1	1:1	41.7/5.3	P02570
20	Actin	1:2	1:2	41.7/5.3	P02570
16	Glucose-regulated protein precursor (Grp78)	1.5: 1	1:1	72.3/5.1	P11021
17	Vimentin	1:1.5	1:2	53.7/5.1	P08670
18	Myosin light chain 1	1:2	1:1.5	16.9/4.6	P16475
19	MRP14	_:3	1:3	13.2/5.7	P06702

Note The marks of "_" present that the spots in the 2D gel stand for "not detectable" for corresponding sample (cited from Journal of Biomechanics, 2010, 43 (12):2339–2347)

In conclusion, our studies demonstrated that the biophysical and immune characteristics of DCs were significantly changed by TMDF, which might be related to the poor immunogenesis of tumors and the less effective immune responses to DCs therapy. The impaired biophysical properties of DCs may be one of many aspects of the immune escape mechanisms of tumors. Clinically, these results indicate the critical importance of the reconstitution of the tumor microenvironment in order to enhance the effectiveness of DCs-based therapy against cancer.

References

1. Soruri A, Zwirner J. Dendritic cells: limited potential in immunotherapy. Int J Biochem Cell Biol. 2005;37(2):241–5.
2. Banchereau J, Palucka AK. Dendritic cells as therapeutic vaccines against cancer. Nat Rev Immunol. 2005;5(4):296–306.
3. Nestle FO, Farkas A, Conrad C. Dendritic-cell-based therapeutic vaccination against cancer. Curr Opin Immunol. 2005;17(2):163–9.
4. Zou W. Immunosuppressive networks in the tumour environment and their therapeutic relevance. Nat Rev Cancer. 2005;5(4):263–74.
5. Steinman RM. The dendritic cell system and its role in immunogenicity. Annu Rev Immunol. 1991;9:271–96.
6. Cella M, Sallusto F, Lanzavecchia A. Origin, maturation and antigen presenting function of dendritic cells. Curr Opin Immunol. 1997;9(1):10–6.
7. Austyn JM. New insights into the mobilization and phagocytic activity of dendritic cells. J Exp Med. 1996;183(4):1287–92.
8. Banchereau J, Steinman RM. Dendritic cells and the control of immunity. Nature. 1998;392 (6673):245–52.
9. Lanzavecchia A. Mechanisms of antigen uptake for presentation. Curr Opin Immunol. 1996;8 (3):348–54.
10. Cumberbatch M, Kimber I. Tumour necrosis factor-alpha is required for accumulation of dendritic cells in draining lymph nodes and for optimal contact sensitization. Immunology. 1995;84(1):31–5.
11. Schuler G, Schuler-Thurner B, Steinman RM. The use of dendritic cells in cancer immunotherapy. Curr Opin Immunol. 2003;15(2):138–47.
12. Melief CJ, et al. Cytotoxic T lymphocyte priming versus cytotoxic T lymphocyte tolerance induction: a delicate balancing act involving dendritic cells. Haematologica, 1999;84 Suppl EHA-4:26–7.
13. Moingeon P. Cancer vaccines. Vaccine. 2001;19(11–12):1305–26.
14. Makarenkova VP, et al. Lung cancer-derived bombesin-like peptides down-regulate the generation and function of human dendritic cells. J Neuroimmunol. 2003;145(1–2):55–67.
15. Makarenkova VP, et al. Lung cancer-derived bombesin-like peptides down-regulate the generation and function of human dendritic cells. J Neuroimmunol. 2003;145(1–2):55–67.
16. Shurin GV, et al. Neuroblastoma-derived gangliosides inhibit dendritic cell generation and function. Cancer Res. 2001;61(1):363–9.
17. Peguet-Navarro J, et al. Gangliosides from human melanoma tumors impair dendritic cell differentiation from monocytes and induce their apoptosis. J Immunol. 2003;170(7):3488–94.
18. Kalinski P, et al. IL-12-deficient dendritic cells, generated in the presence of prostaglandin E2, promote type 2 cytokine production in maturing human naive T helper cells. J Immunol. 1997;159(1):28–35.

19. Gabrilovich DI, et al. Decreased antigen presentation by dendritic cells in patients with breast cancer. Clin Cancer Res. 1997;3(3):483–90.
20. Allavena P, et al. IL-10 prevents the differentiation of monocytes to dendritic cells but promotes their maturation to macrophages. Eur J Immunol. 1998;28(1):359–69.
21. D'Orazio TJ, Niederkorn JY. A novel role for TGF-beta and IL-10 in the induction of immune privilege. J Immunol. 1998;160(5):2089–98.
22. Almand B, et al. Increased production of immature myeloid cells in cancer patients: a mechanism of immunosuppression in cancer. J Immunol. 2001;166(1):678–89.
23. McBride JM, et al. IL-10 alters DC function via modulation of cell surface molecules resulting in impaired T-cell responses. Cell Immunol. 2002;215(2):162–72.
24. Tsujitani S, et al. Infiltration of dendritic cells in relation to tumor invasion and lymph node metastasis in human gastric cancer. Cancer. 1990;66(9):2012–6.
25. Becker Y. Anticancer role of dendritic cells (DC) in human and experimental cancers—a review. Anticancer Res. 1992;12(2):511–20.
26. Shurin GV, et al. Loss of new chemokine CXCL14 in tumor tissue is associated with low infiltration by dendritic cells (DC), while restoration of human CXCL14 expression in tumor cells causes attraction of DC both in vitro and in vivo. J Immunol. 2005;174(9):5490–8.
27. Gabrilovich DI, et al. Defects in the function of dendritic cells in murine retroviral infection. Adv Exp Med Biol. 1995;378:469–72.
28. Thurnher M, et al. Human renal-cell carcinoma tissue contains dendritic cells. Int J Cancer. 1996;68(1):1–7.
29. Garrity T, et al. Increased presence of CD34+ cells in the peripheral blood of head and neck cancer patients and their differentiation into dendritic cells. Int J Cancer. 1997;73(5):663–9.
30. Ninomiya T, et al. Dendritic cells with immature phenotype and defective function in the peripheral blood from patients with hepatocellular carcinoma. J Hepatol. 1999;31(2):323–31.
31. Schwaab T, et al. In vivo description of dendritic cells in human renal cell carcinoma. J Urol. 1999;162(2):567–73.
32. Tourkova IL, et al. Mechanisms of dendritic cell-induced T cell proliferation in the primary MLR assay. Immunol Lett. 2001;78(2):75–82.
33. Benvenuti F, et al. Requirement of Rac1 and Rac2 expression by mature dendritic cells for T cell priming. Science. 2004;305(5687):1150–3.
34. Mempel TR, Henrickson SE, Von Andrian UH. T-cell priming by dendritic cells in lymph nodes occurs in three distinct phases. Nature. 2004;427(6970):154–9.
35. Stoll S, et al. Dynamic imaging of T cell-dendritic cell interactions in lymph nodes. Science. 2002;296(5574):1873–6.
36. Cavanagh LL, Weninger W. Dendritic cell behaviour in vivo: lessons learned from intravital two-photon microscopy. Immunol Cell Biol. 2008;86(5):428–38.
37. Alvarez D, Vollmann EH, von Andrian UH. Mechanisms and consequences of dendritic cell migration. Immunity. 2008;29(3):325–42.
38. Jiang Y, et al. Adhesion of monocyte-derived dendritic cells to human umbilical vein endothelial cells in flow field decreases upon maturation. Clin Hemorheol Microcirc. 2005;32 (4):261–8.
39. Zeng Z, et al. Biophysical studies on the differentiation of human CD14+ monocytes into dendritic cells. Cell Biochem Biophys. 2006;45(1):19–30.
40. Lee WC, et al. Functional impairment of dendritic cells caused by murine hepatocellular carcinoma. J Clin Immunol. 2004;24(2):145–54.
41. Satthaporn S, et al. Dendritic cells are dysfunctional in patients with operable breast cancer. Cancer Immunol Immunother. 2004;53(6):510–8.
42. Le Gal JM, et al. Conformational changes in membrane proteins of multidrug-resistant K562 and primary rat hepatocyte cultures as studied by Fourier transform infrared spectroscopy. Anticancer Res. 1994;14(4A):1541–8.
43. Shinitzky M. Biomembranes. Weinheim: VCH. 1; 1993.
44. Nathan I, et al. Alterations in membrane lipid dynamics of leukemic cells undergoing growth arrest and differentiation: dependency on the inducing agent. Exp Cell Res. 1998;239(2):442–6.

45. Dustin ML, Cooper JA. The immunological synapse and the actin cytoskeleton: molecular hardware for T cell signaling. Nat Immunol. 2000;1(1):23–9.
46. Pantaloni D, Le Clainche C, Carlier MF. Mechanism of actin-based motility. Science. 2001;292(5521):1502–6.
47. Burns S, et al. Maturation of DC is associated with changes in motile characteristics and adherence. Cell Motil Cytoskeleton. 2004;57(2):118–32.
48. Ramesh J, et al. Application of FTIR microscopy for the characterization of malignancy: H-ras transfected murine fibroblasts as an example. J Biochem Biophys Methods. 2001;50(1):33–42.
49. Byrne SN, Halliday GM. Dendritic cells: making progress with tumour regression? Immunol Cell Biol. 2002;80(6):520–30.
50. Sozzani S. Dendritic cell trafficking: more than just chemokines. Cytokine Growth Factor Rev. 2005;16(6):581–92.
51. de Vries IJ, et al. Maturation of dendritic cells is a prerequisite for inducing immune responses in advanced melanoma patients. Clin Cancer Res. 2003;9(14):5091–100.
52. Randolph GJ, Angeli V, Swartz MA. Dendritic-cell trafficking to lymph nodes through lymphatic vessels. Nat Rev Immunol. 2005;5(8):617–28.
53. Karin M, Greten FR. NF-kappaB: linking inflammation and immunity to cancer development and progression. Nat Rev Immunol. 2005;5(10):749–59.
54. Sun B, Karin M. NF-kappaB signaling, liver disease and hepatoprotective agents. Oncogene. 2008;27(48):6228–44.
55. Burkly L, et al. Expression of relB is required for the development of thymic medulla and dendritic cells. Nature. 1995;373(6514):531–6.
56. Bao B, et al. The immunological contribution of NF-kappaB within the tumor microenvironment: a potential protective role of zinc as an anti-tumor agent. Biochim Biophys Acta. 2012;1825(2):160–72.
57. Kuwabara T, et al. CCR7 ligands up-regulate IL-23 through PI3-kinase and NF-kappa B pathway in dendritic cells. J Leukoc Biol. 2012;92(2):309–18.
58. Shih VF, et al. Control of RelB during dendritic cell activation integrates canonical and noncanonical NF-kappaB pathways. Nat Immunol. 2012;13(12):1162–70.
59. Luo L, et al. Effects of tolerogenic dendritic cells generated by siRNA-mediated RelB silencing on immune defense and surveillance functions of T cells. Cell Immunol. 2013;282(1):28–37.
60. Rabinovich GA, Gabrilovich D, Sotomayor EM. Immunosuppressive strategies that are mediated by tumor cells. Annu Rev Immunol. 2007;25:267–96.
61. Steinman RM. Dendritic cells in vivo: a key target for a new vaccine science. Immunity. 2008;29(3):319–24.
62. Steinman RM. Decisions about dendritic cells: past, present, and future. Annu Rev Immunol. 2012;30:1–22.
63. Monti P, et al. Tumor-derived MUC1 mucins interact with differentiating monocytes and induce IL-10^{high}IL-12^{low} regulatory dendritic cell. J Immunol. 2004;172(12):7341–9.
64. Carlsson L, et al. Actin polymerizability is influenced by profilin, a low molecular weight protein in non-muscle cells. J Mol Biol. 1977;115(3):465–83.
65. Didry D, Carlier MF, Pantaloni D. Synergy between actin depolymerizing factor/cofilin and profilin in increasing actin filament turnover. J Biol Chem. 1998;273(40):25602–11.
66. Dong R, et al. Dendritic cells from CML patients have altered actin organization, reduced antigen processing, and impaired migration. Blood. 2003;101(9):3560–7.
67. Himelstein BP, et al. Metalloproteinases in tumor progression: the contribution of MMP-9. Invasion Metastasis. 1994;14(1–6):246–58.
68. Nawrocki B, et al. Expression of matrix metalloproteinases and their inhibitors in human bronchopulmonary carcinomas: quantitative and morphological analyses. Int J Cancer. 1997;72(4):556–64.

Chapter 5
Effects of Traditional Chinese Medicine on DCs Under Tumor Microenvironment

Abstract Tumor microenvironment-derived factors (TMDF) can impair the immune functions of DCs through the disturbance of biophysical characteristics. Reconstruction of the tumor microenvironment could improve the clinical effectiveness of DCs-based immune therapy against cancer. The Gekko sulfated polysaccharide-protein complex (GSPP), a kind of medicine isolated from traditional Chinese herbs Gekko swinhonis Guenther, could partially or completely improve the biophysical characteristics of DCs in the tumor microenvironment. The secretion of IL-12 did not change compared with that of DCs in tumor microenvironment, but the secretion of IL-10 was resumed to control level, indicating that GSPP could partially restore the defective biophysical characteristics of DCs via modifying the HCC microenvironment and decreasing the secretion of IL-10 of DCs.

Keywords DCs · Biophysical characteristics · Tumor microenvironment-derived factors · Traditional Chinese medicine

Immune cells that participated in the elimination of tumor cells include innate effectors such as gamma delta T cells, natural killer cells, DCs, natural killer T cells, and members of adaptive immune cells (B and T cells) [1–6]. The activation of T cells, B cells, natural killer cells, and natural killer T cells and the secretion of their effect cytokines are related with DCs [7, 8]. DCs are professional antigen presenting cells (APC) and they are in charge of integrating a mass of incoming signals and orchestrate the immune response [9]. This kind of APC is involves in innate and adaptive immunity and regulate immunity through inducing antigen-specific unresponsiveness of lymphocytes by deleting effector cells and inducting regulatory T cells [10, 11]. DCs also play an important role in the antitumor process [12, 13]. The amount, biophysical properties, and functions of DCs have a very close relationship with the occurrence, development, and prognosis of tumor, which is of vital significance for us to study how tumors escape from the immune surveillance and how to select an effective method to treat tumors.

© The Author(s) 2015
Z. Zeng et al., *Dendritic Cells: Biophysics, Tumor Microenvironment and Chinese Traditional Medicine*, SpringerBriefs in Biochemistry and Molecular Biology, DOI 10.1007/978-94-017-7405-5_5

DCs' biophysical characteristics, such as cell migration, cytoskeleton, membrane fluidity, intracellular calcium concentration, cell osmotic fragility, and cell viscoelastic property play an important role in immune response. DCs' uptake antigens in peripheral tissues and then some biophysical characteristics, which may include deformability, migration, and membrane fluidity, are beginning to play out. DCs migrate to peripheral lymphoid organs, presenting antigens to primary T cells and stimulating the activation of T cells. Cytoskeleton of DCs has an important status, which guides the direction of migration and provides the momentum. There are three types of cytoskeleton proteins: microtubules, intermediate filaments, and actin filaments [14]. The migration of DCs mainly depends on the reorganization and distribution of actin filaments and microtubules [15], which is primarily through some pathways, for example, cell polarization, the formation of pseudopodia, etc. Then protrusions adhere to the substrate and form focal adhesions, and the cell body moves toward and drawing on the tail. Through these complex processes, the migration process was able to be completed. Visibly, the biophysical properties of DCs, such as deformation and migration, provide the effective basis for DCs' immunological functions.

Tumor microenvironment plays a key role in tumor development [16–19]. Recent studies suggest that some tumors may avoid immune destruction via inhibiting DCs' function, including impaired DCs migration, inhibition of DCs cross-presentation functions, inhibition of DCs maturation, and so on. This may lead to immune tolerance [20]. Some reports have shown that the tumor microenvironment of HCC cells could affect the biophysical characteristics of DCs [21, 22]. The results show that coculture of DCs with HCC cell line would decrease abilities both of imDCs and mDCs. The tumor microenvironment of HCCs decreased the membrane fluidity of DCs, reduced charges on the membrane, and increased its osmotic fragilities and viscoelastic property [22].

It is reported that tumor microenvironment-derived factors (TMDF), for example, cytokines, could interfere with biophysical properties of DCs [23]. Our results have shown that the exposure to supernatant from normal human liver cell line HL-7702 has no effect on the migration capability and configuration of the cytoskeleton of DCs, which was consistent in previous studies using several types of normal cell lines [24, 25]. However, hepatocellular carcinoma cells SMMC-7721 conditioned medium could signally change the biophysical behaviors of DCs, including declined cell deformability, migration and electrophoresis mobility, augmented osmotic fragilities, and changed organizations of cytoskeleton proteins [21]. These results indicate that the SMMC-7721 conditioned medium contains some kinds of soluble materials that have direct impact on DCs and deactivates the immune functions of DCs. DCs cultured in the SMMC-7721 conditioned medium had poor deformability, which makes the DCs easily destructible due to blood flow and difficult to take place transendothelial migration. Charges on the surface of DCs also decreased, so that DCs' affinity with other cells and/or extracellular matrix is also decreased. This may have effect on cytokines, which induces an accurate migration of DCs [26, 27]. Confocal microscopy showed that the SMMC-7721 conditioned medium could influence cytoskeleton protein of DCs. The F-actin and tubulin are much shorter than normal. It is reported that the cytoskeleton serves as infrastructure of cell migration, deformation, and signaling [28, 29]. Hence, the changes in cytoskeleton in DCs could impair the migration, deformation, and signaling function of DCs.

Some traditional Chinese medicine (TCM) have been shown to be able to improve the biophysical function of DCs under tumor microenvironment. Gekko swinhonis Guenther, commonly known as Gecko, is an animal of the genus Gekko, the family of Gekkonidae, which is widely spread in the north and central regions of China. The dried whole body has been used as an anticancer drug in TCM for hundreds of years [30]. Gekko sulfated polysaccharide-protein complex (GSPP) was isolated from Gekko swinhonis Guenther. It was chemically characterized as a sulfated polysaccharide-protein complex with O-glycopeptides linkages. GSPP has been reported to not only have a direct inhibitory effect on hepatocellular carcinoma cells, but also improve DCs' function under tumor microenvironments [30].

The results of certain trials by some researchers showed that GSPP had direct antitumor activity. GSPP inhibited the proliferation of SMMC-7721 cells and blocked cells in the S phase. Direct toxicity against cells was not observed. Furthermore, GSPP inhibited the migration of SMMC-7721 cells with the reduction of intracellular calcium. Actin filaments were polymerized and accumulated in the cytoplasm of the treated cells, which was mediated by calcium ion, and increasing the expression of β-actin RNA [30, 31]. It was also reported that GSPP may block BEL-7402 cells in G_2/M phase [30].

GSPP could improve the biophysical function of DCs under tumor microenvironment, which has been shown in some further researches. When cultured in SMMC + GSPP conditioned medium, the migration function of DCs recovered, and also increased in the electrophoresis mobility, the negative charge on DCs' surface, and the length and density of DCs' protrusions. Results of micropipette aspiration experiment showed that the deformability of DCs was partially recovered after it was cultured in SMMC + GSPP conditioned medium (Fig. 5.1). The controlled trials cultured DCs in four kinds of conditioned mediums, including control, HL, SMMC, and SMMC + GSPP. The data indicated that the deformability of DCs in

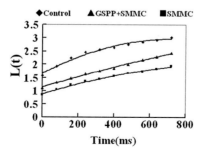

Fig. 5.1 The deformation of DCs under SMMC-7721 cell microenvironment as measured by micropipette aspiration. The length in the micropipette, L(t), is a function of time (cited from Cell Biochem Biophys. 2012;62(1):193–201. P196)

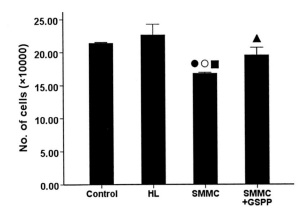

Fig. 5.2 Cells that had penetrated through the chamber were counted using hemocytometer. $p = 0.001$. *Filled circle* versus control group, $p = 0.001$, versus HL group, $p < 0.001$, *Filled square* versus SMMC + GSPP group, $p = 0.01$, *Filled triangle* versus HL group, $p = 0.005$ (cited from Cell Biochem Biophys. 2012;62(1):193–201. P196)

SMMC + GSPP conditioned medium became better than that of SMMC, even though it was still less than that of control cells.

The migration capability of DCs was also reversed. The results of transwell assay demonstrated that DCs cultured in SMMC + GSPP conditioned medium had more migrating cells than those of SMMC, which was also basically the same number of migrating cells with the control cells (Fig. 5.2).

Researchers also found that the electrophoresis mobility of DCs was improved. DCs cultured in SMMC + GSPP conditioned medium were faster than those cultured in SMMC conditioned medium, although it was slower than that of control cells (Fig. 5.3).

SMMC + GSPP conditioned medium changed the organizations of cytoskeleton proteins of DCs. While observing DCs by confocal microscope, results indicated that DCs cultured in SMMC + GSPP conditioned medium had longer length and greater density compared with DCs cultured in SMMC conditioned medium [21] (Figs. 5.4 and 5.5).

Some materials, such as VEGF and IL-10, which are presented in the tumor microenvironment, were capable of altering the bio-rheological behaviors of DCs. These cytokines played key roles in inhibiting the differentiation and maturation of DCs [32, 33]. IL-10 has already been proved a factor that participates in immune tolerance in the recent years [34–37]. Nevertheless, there was no difference between SMMC-7721 and GSPP-treated SMMC-7721 cells in the levels of VEGF and IL-10 secretion. As reported by many researchers, the type of cytokines, especially cytokines IL-12 and IL-10, determined the type of immune responses induced by DCs. IL-10 tended to trigger an immunological drift, which changed Th_1 to Th_2, and this resulted in a failure to lyse tumor cells [38–40]. In addition, IL-10 was one of the factors responsible for inducing the immunological tolerance of DCs.

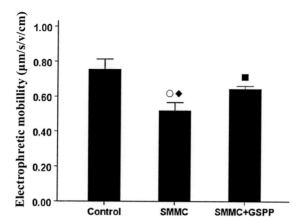

Fig. 5.3 Electrophoretic mobilities of DCs under different conditions. $p = 0.002$. The values represent means ±SD. versus control group, $p = 0.001$; *Filled diamonds* versus SMMC + GSPP group, $p = 0.014$; *Filled square* versus control group, $p = 0.021$ (cited from Cell Biochem Biophys. 2012;62(1):193–201. P197)

Fig. 5.4 Tubulin sectional images of DCs under different conditions (600×). **a** Control group. **b** HL group. **c** SMMC group. **d** SMMC+GSPP group (cited from Cell Biochem Biophys. 2012;62 (1):193–201. P198)

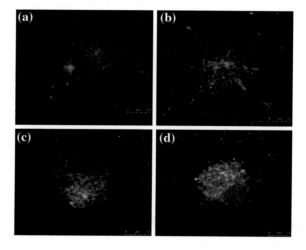

The pathways may decrease the expression of costimulatory molecules, impaired T-cell responses, and induced T-cell anergy [36, 37, 41]. Similarly, some researches showed that IL-10 inhibitors could reverse the dysfunction of DCs in the conditions of tumors [42–45]. Our research indicated that when cultured in SMMC conditioned medium, DCs can produce more IL-10 but less IL-12 than those in control medium (Figs. 5.4 and 5.5). This altered cytokine profile may be of great help to initiate Th$_2$ response rather than Th$_1$ response. SMMC + GSPP conditioned mediums had no effect on IL-12 secretion, whereas it significantly inhibited the secretion of IL-10 of DCs, which could help to initiate the cellular immune response

Fig. 5.5 F-actin in DC under different conditions. Cells were incubated with the phalloidin *rhodamine* and then photographed under oil microscope (600×). The serial sectional images (*left panels*) were reconstructed into three-dimensional images (*right panels*). **a** Control group. **b** HL group. **c** SMMC group. **d** SMMC + GSPP group (Cell Biochem Biophys. 2012, 62(1):193–201)

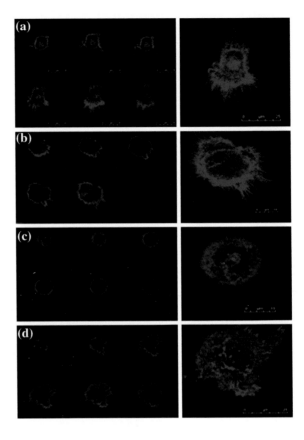

of tumor-specific antigen. Furthermore, the detected IL-10 must be produced by DCs, as the production of IL-10 detected from the supernatants of SMMC-7721 and GSPP-treated SMMC-7721 cells were low. Consistent with other reports, our research suggested that some factors present in the tumor microenvironment mediated DCs' dysfunction partly through inducing the secretion of IL-10 in DCs [46, 47]. However regretfully, GSPP would not change the osmotic fragility dysfunction of DCs under tumor microenvironment [21] (Figs. 5.6 and 5.7).

Taken together, immune suppressive pathways engaged by cancer cells during tumor progression may be varied. For example, the tumor microenvironment-derived suppressive cytokines can impair motility and immune response of DCs through the disturbance of biophysical characteristics and reorganization of cytoskeletons. GSPP, a kind of medicine isolated from traditional Chinese herbs, has a direct inhibiting effect on tumor cells. GSPP also could improve the biophysical characteristics of DCs under tumor microenvironments. This is important in limiting the growth of tumors, even to the extent that it shrinks the tumor and is a guidance for tumor therapy.

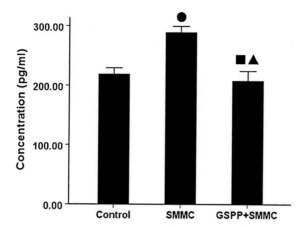

Fig. 5.6 IL-10 secretion of DCs under different conditions. *p* < 0.001. *Filled circle* versus control group, *p* = 0.001, *Filled square* versus control group, *p* = 0.33, *Filled triangle* versus SMMC group, *p* < 0.001 (cited from Cell Biochem Biophys. 2012; 62(1): 193–201. P198)

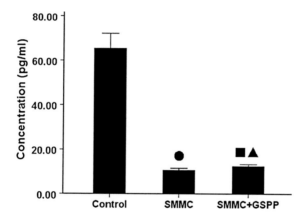

Fig. 5.7 IL-12 secretion of DCs under different conditions. *p* < 0.001. *Filled circle* versus control group, *p* = 0.013, *Filled square* versus control group, *p* = 0.014, *Filled triangle* versus SMMC group, *p* = 0.208 (cited from Cell Biochem Biophys. 2012;62(1):193–201)

References

1. Gao Y, et al. Gamma delta T cells provide an early source of interferon gamma in tumor immunity. J Exp Med. 2003;198(3):433–42.
2. Borg C, et al. Novel mode of action of c-kit tyrosine kinase inhibitors leading to NK cell-dependent antitumor effects. J Clin Invest. 2004;114(3):379–88.
3. Smyth MJ, Godfrey DI, Trapani JA. A fresh look at tumor immunosurveillance and immunotherapy. Nat Immunol. 2001;2(4):293–9.

4. Hildner K, et al. Batf3 deficiency reveals a critical role for CD8alpha+ dendritic cells in cytotoxic T cell immunity. Science. 2008;322(5904):1097–100.
5. Crowe NY, Smyth MJ, Godfrey DI. A critical role for natural killer T cells in immunosurveillance of methylcholanthrene-induced sarcomas. J Exp Med. 2002;196 (1):119–27.
6. Shankaran V, et al. IFNgamma and lymphocytes prevent primary tumour development and shape tumour immunogenicity. Nature. 2001;410(6832):1107–11.
7. Banchereau J, Steinman RM. Dendritic cells and the control of immunity. Nature. 1998;392(6673):245–52.
8. Shortman K, Liu YJ. Mouse and human dendritic cell subtypes. Nat Rev Immunol. 2002;2 (3):151–61.
9. Banchereau J, et al. Immunobiology of dendritic cells. Annu Rev Immunol. 2000;18:767–811.
10. Pletinckx K, et al. Role of dendritic cell maturity/costimulation for generation, homeostasis, and suppressive activity of regulatory T cells. Front Immunol. 2011;2:39.
11. Gilliet M, Liu YJ. Human plasmacytoid-derived dendritic cells and the induction of T-regulatory cells. Hum Immunol. 2002;63(12):1149–55.
12. Zeid NA, Muller HK. S100 positive dendritic cells in human lung tumors associated with cell differentiation and enhanced survival. Pathology. 1993;25(4):338–43.
13. Lespagnard L, et al. Tumor-infiltrating dendritic cells in adenocarcinomas of the breast: a study of 143 neoplasms with a correlation to usual prognostic factors and to clinical outcome. Int J Cancer. 1999;84(3):309–14.
14. Jiang P, Enomoto A, Takahashi M. Cell biology of the movement of breast cancer cells: intracellular signalling and the actin cytoskeleton. Cancer Lett. 2009;284(2):122–30.
15. Jaworski J, et al. Dynamic microtubules regulate dendritic spine morphology and synaptic plasticity. Neuron. 2009;61(1):85–100.
16. DeNardo DG, Coussens LM. Inflammation and breast cancer. Balancing immune response: crosstalk between adaptive and innate immune cells during breast cancer progression. Breast Cancer Res. 2007;9(4):212.
17. Talmadge JE, Donkor M, Scholar E. Inflammatory cell infiltration of tumors: Jekyll or Hyde. Cancer Metastasis Rev. 2007;26(3–4):373–400.
18. Waldner M, Schimanski CC, Neurath MF. Colon cancer and the immune system: the role of tumor invading T cells. World J Gastroenterol. 2006;12(45):7233–8.
19. Conejo-Garcia JR, et al. Letal, A tumor-associated NKG2D immunoreceptor ligand, induces activation and expansion of effector immune cells. Cancer Biol Ther. 2003;2(4):446–51.
20. Apetoh L, et al. Harnessing dendritic cells in cancer. Semin Immunol. 2011;23(1):42–9.
21. Chen D, et al. Effects of Gekko sulfated polysaccharide-protein complex on the defective biorheological characters of dendritic cells under tumor microenvironment. Cell Biochem Biophys. 2012;62(1):193–201.
22. Zeng Z, et al. Hepatocellular carcinoma cells deteriorate the biophysical properties of dendritic cells. Cell Biochem Biophys. 2009;55(1):33–43.
23. Zeng Z, et al. Tumor-derived factors impaired motility and immune functions of dendritic cells through derangement of biophysical characteristics and reorganization of cytoskeleton. Cell Motil Cytoskeleton. 2007;64(3):186–98.
24. Li L, et al. Hepatoma cells inhibit the differentiation and maturation of dendritic cells and increase the production of regulatory T cells. Immunol Lett. 2007;114(1):38–45.
25. Kiertscher SM, et al. Tumors promote altered maturation and early apoptosis of monocyte-derived dendritic cells. J Immunol. 2000;164(3):1269–76.
26. Stoll S, et al. Dynamic imaging of T cell-dendritic cell interactions in lymph nodes. Science. 2002;296(5574):1873–6.
27. Randolph GJ, Angeli V, Swartz MA. Dendritic-cell trafficking to lymph nodes through lymphatic vessels. Nat Rev Immunol. 2005;5(8):617–28.
28. Vicente-Manzanares M, Sanchez-Madrid F. Role of the cytoskeleton during leukocyte responses. Nat Rev Immunol. 2004;4(2):110–22.

29. Beemiller P, Krummel MF. Mediation of T-cell activation by actin meshworks. Cold Spring Harb Perspect Biol. 2010;2(9):a002444.

30. Wu X, Chen D, Xie GR. Effects of Gekko sulfated polysaccharide on the proliferation and differentiation of hepatic cancer cell line. Cell Biol Int. 2006;30(8):659–64.

31. Chen D, et al. Effects of Gekko sulfated polysaccharide-protein complex on human hepatoma SMMC-7721 cells: inhibition of proliferation and migration. J Ethnopharmacol. 2010;127(3):702–8.

32. Strauss L, et al. Dual role of VEGF family members in the pathogenesis of head and neck cancer (HNSCC): possible link between angiogenesis and immune tolerance. Med Sci Monit. 2005;11(8): BR280–292.

33. Yang AS, Lattime EC. Tumor-induced interleukin 10 suppresses the ability of splenic dendritic cells to stimulate CD4 and CD8 T-cell responses. Cancer Res. 2003;63(9):2150–7.

34. Li L, et al. Hepatoma cells inhibit the differentiation and maturation of dendritic cells and increase the production of regulatory T cells. Immunol Lett. 2007;114(1):38–45.

35. Li X, et al. Induction of type 2 T helper cell allergen tolerance by IL-10-differentiated regulatory dendritic cells. Am J Respir Cell Mol Biol. 2010;42(2):190–9.

36. McBride JM, et al. IL-10 alters DC function via modulation of cell surface molecules resulting in impaired T-cell responses. Cell Immunol. 2002;215(2):162–72.

37. Caux C, et al. Interleukin 10 inhibits T cell alloreaction induced by human dendritic cells. Int Immunol. 1994;6(8):1177–85.

38. Hilkens CM, et al. Human dendritic cells require exogenous interleukin-12-inducing factors to direct the development of naive T-helper cells toward the Th1 phenotype. Blood. 1997;90(5):1920–6.

39. Steinbrink K, et al. Interleukin-10-treated human dendritic cells induce a melanoma-antigen-specific anergy in CD8$^+$ T cells resulting in a failure to lyse tumor cells. Blood. 1999;93(5):1634–42.

40. Steinbrink K, et al. CD4$^+$ and CD8$^+$ anergic T cells induced by interleukin-10-treated human dendritic cells display antigen-specific suppressor activity. Blood. 2002;99(7):2468–76.

41. Li X, et al. Induction of type 2 T helper cell allergen tolerance by IL-10-differentiated regulatory dendritic cells. Am J Respir Cell Mol Biol. 2010;42(2):190–9.

42. Vicari AP, et al. Reversal of tumor-induced dendritic cell paralysis by CpG immunostimulatory oligonucleotide and anti-interleukin 10 receptor antibody. J Exp Med. 2002;196(4):541–9.

43. Brignole C, et al. Anti-IL-10R antibody improves the therapeutic efficacy of targeted liposomal oligonucleotides. J Control Release. 2009;138(2):122–7.

44. Singh A, et al. An injectable synthetic immune-priming center mediates efficient T-cell class switching and T-helper 1 response against B cell lymphoma. J Control Release. 2011;155(2):184–92.

45. Singh A, Suri S, Roy K. In-situ crosslinking hydrogels for combinatorial delivery of chemokines and siRNA-DNA carrying microparticles to dendritic cells. Biomaterials. 2009;30(28):5187–200.

46. Kuo PL, et al. Lung cancer-derived galectin-1 mediates dendritic cell anergy through inhibitor of DNA binding 3/IL-10 signaling pathway. J Immunol. 2011;186(3):1521–30.

47. Zhao F, et al. Activation of p38 mitogen-activated protein kinase drives dendritic cells to become tolerogenic in ret transgenic mice spontaneously developing melanoma. Clin Cancer Res. 2009;15(13):4382–90.

Chapter 6
Future Perspectives

Abstract The basic researches and clinical applications of DCs have achieved great progress, but there are still many challenges that need to be overcome. The factors that affect DC functions, e.g., chemical and physical factors, are focused from the interdisciplinary viewpoint. Here, the future research directions are summarized.

Keywords Dentritic cells · Immune response · DCs-based therapies · Vaccine

DCs comprise a population of leukocytes with the capability of activating specific immune responses to promote immunity or induce tolerance, which capture, process, and present antigens thereby activating T cells that carry cognate receptors for these presented antigens. Consequently, DCs serve a vital function in initiating adaptive immunity and orchestrating the immune response outcome. The motility of DCs is important for migration of imDCs in peripheral tissues and physical interaction between mDCs and naive T cells in secondary lymph nodes. The tumor microenvironment-derived suppressive factors can exert undesirable effects on DCs by either inhibiting maturation and then skewing them unable to promote specific immune response, or transforming them into immunosuppressive cells capable of inducing regulatory T cells collectively creating significant obstacles and challenges in cancer immunotherapy. Interestingly, the combination of a vaccine with immunomodulatory molecules to neutralize inhibitory signals may be necessary to produce effective T-cell immune response. Ample evidences support the feasibility to overcome the immune paralysis of cancer-associated DCs, such as improvement of immunosuppressive tumor environment, proper vehicles to deliver antigens directly to DCs, effective receptors or specific DC subsets targeted to induce humoral and cellular responses. Therefore, the reprogramming of cancer-associated DCs will be an attractive field; DC responses to specific signals are crucial for understanding their immunological behaviors and improvement of clinical efficiency of DCs-based therapy against cancer. On the one hand, the future work will be focused on the existence of DC subsets with specialized functions, the impact of the antigen intracellular trafficking on cross-presentation, the influence of

© The Author(s) 2015
Z. Zeng et al., *Dendritic Cells: Biophysics, Tumor Microenvironment
and Chinese Traditional Medicine*, SpringerBriefs in Biochemistry
and Molecular Biology, DOI 10.1007/978-94-017-7405-5_6

maturation signals received by DCs on the outcome of the immune response, and new antigen-loaded strategies and migratory capability of reprogrammed DCs. On the other hand, except for chemical factors (e.g., cytokines), how the physical factors (e.g., physiological flow-derived shear stress) affect the immune function of DCs will also be a non-neglectable aspect. Thus, the interdisciplinary knowledge which is related to the tumor microenvironment, motilities and immune regulatory function of DCs from the viewpoint of tumor immunology, biophysics, mechanobiology, cell biology, and traditional Chinese medicine will promote the discovery of new strategies for transforming cancer-associated DCs into effective antigen-presenting cells and exploring DCs-based immunotherapy. Briefly, the better personalized DCs-based antitumor vaccination in situ or in vivo will be a promising immunotherapy.

In conclusion, in spite of a number of challenges, the present is viewed as an exciting time to study vaccine development and foresee that continuing to design DC-based therapies will allow us to prevent and treat many of the major illnesses for which no effective vaccine currently exists.

Acknowledgments These works were supported by the National Natural Science Foundations of China (No. 11162003 and 31260227), Key Project of Chinese Ministry of Education (210196), Supporting Project of Distinguished Young Science and Technology Scholars of Guizhou Province (2011-24), Offends Pass Item of Science and Technology for Social Development in Guizhou Province (GZ-SY-[2011]-3065), Special Foundation of Governor of Guizhou Province (2009-79), Foundation of Science and Technology of Guizhou Province (J-2008-2274, J-2013-2058), Technology Foundation for Selected Overseas Chinese of Guizhou Province (2013-8), Scientific and Technological Innovation Talent Team of Guizhou Province (2015–4021).